Advances in Intelligent Systems and Computing

Volume 1061

The series "Advances in Intelligent Systems and Computing" contains publications on theory, applications, and design methods of Intelligent Systems and Intelligent Computing. Virtually all disciplines such as engineering, natural sciences, computer and information science, ICT, economics, business, e-commerce, environment, healthcare, life science are covered. The list of topics spans all the areas of modern intelligent systems and computing such as: computational intelligence, soft computing including neural networks, fuzzy systems, evolutionary computing and the fusion of these paradigms, social intelligence, ambient intelligence, computational neuroscience, artificial life, virtual worlds and society, cognitive science and systems, Perception and Vision, DNA and immune based systems, self-organizing and adaptive systems, e-Learning and teaching, human-centered and human-centric computing, recommender systems, intelligent control, robotics and mechatronics including human-machine teaming, knowledge-based paradigms, learning paradigms, machine ethics, intelligent data analysis, knowledge management, intelligent agents, intelligent decision making and support, intelligent network security, trust management, interactive entertainment, Web intelligence and multimedia.

The publications within "Advances in Intelligent Systems and Computing" are primarily proceedings of important conferences, symposia and congresses. They cover significant recent developments in the field, both of a foundational and applicable character. An important characteristic feature of the series is the short publication time and world-wide distribution. This permits a rapid and broad dissemination of research results.

** Indexing: The books of this series are submitted to ISI Proceedings, EI-Compendex, DBLP, SCOPUS, Google Scholar and Springerlink **

More information about this series at http://www.springer.com/series/11156

Aleksandra Gruca · Tadeusz Czachórski ·
Sebastian Deorowicz · Katarzyna Harężlak ·
Agnieszka Piotrowska
Editors

Man-Machine Interactions 6

6th International Conference on
Man-Machine Interactions, ICMMI 2019,
Cracow, Poland, October 2–3, 2019

 Springer

Editors
Aleksandra Gruca
Institute of Informatics
Silesian University of Technology
Gliwice, Poland

Sebastian Deorowicz
Institute of Informatics
Silesian University of Technology
Gliwice, Poland

Agnieszka Piotrowska
Institute of Informatics
Silesian University of Technology
Gliwice, Poland

Tadeusz Czachórski
Institute of Theoretical and Applied
Informatics
Polish Academy of Sciences
Gliwice, Poland

Katarzyna Harężlak
Institute of Informatics
Silesian University of Technology
Gliwice, Poland

ISSN 2194-5357   ISSN 2194-5365   (electronic)
Advances in Intelligent Systems and Computing
ISBN 978-3-030-31963-2   ISBN 978-3-030-31964-9   (eBook)
https://doi.org/10.1007/978-3-030-31964-9

This Springer imprint is published by the registered company Springer Nature Switzerland AG
The registered company address is: Gewerbestrasse 11, 6330 Cham, Switzerland

The computer is incredibly fast, accurate, and stupid. Man is unbelievably slow, inaccurate, and brilliant. The marriage of the two is a force beyond calculation.

Leo Chernie

Preface

This volume contains the proceedings of the 6th International Conference on Man-Machine Interactions, ICMMI 2019, which was held in Kraków, Poland, 2–3 October 2019. Since 2009, the biennial ICMMI Conference has been organized jointly by the Institute of Informatics at the Silesian University of Technology and the Institute of Theoretical and Applied Informatics of the Polish Academy of Sciences in Gliwice, Poland. The first ICMMI Conference was dedicated to the memory of Adam Mrózek, a distinguished scientist in the area of decision support systems in industrial applications. This year, we are celebrating the tenth anniversary of ICMMI with the sixth conference of the series, which has become a regular event in the calendar of scientific meetings worldwide.

The ICMMI Conference constitutes an international forum bringing together academic and industrial researchers interested in all aspects of theory and practice of man–machine interactions. Over the last ten years, thousands of scientists from all over the world have attended to discuss their latest research, exchange ideas, and learn about new developments in the field. Nowadays, intelligent machines appear to influence nearly every aspect of modern society by improving efficiencies and augmenting our human capabilities. The topic of human–computer interactions has thus become more important and more complex than ever.

Human–computer interaction is a multidisciplinary field of study focusing on the design of computer technology and, in particular, the interaction between humans (the users) and computers. For the last decades, this field expanded from its initial focus on individual and generic user behaviour to the widest possible spectrum of human experiences and activities. In summary, research in this field covers various technological aspects related to understanding and improving the ways in which people communicate with machines.

The ICMMI 2019 Conference attracted authors from 11 different countries across the world who submitted contributed papers presenting a broad range of topics related to man–machine interactions. The review process was carried out by a multidisciplinary team of 84 members of the Programme Committee with the help of the external reviewers. Each paper was subjected to at least two independent reviews. After the peer-review process, 24 high-quality papers were selected for publication in the ICMMI 2019 proceedings. These papers were divided into the sections 'human–computer interfaces', 'artificial intelligence and knowledge discovery', 'pattern recognition', 'bio-data and bio-signal analysis', and 'algorithms, optimization, and signal processing'. We here express our deepest gratitude to all members of the Programme Committee and the associate reviewers for their time, effort, and invaluable contribution to the success of the conference, ensuring that the excellence of the scientific programme is maintained.

Special thanks to our distinguished keynote speakers who enriched the conference with their inspiring talks presenting the latest advances and developments in the field of man–machine interactions.

We would also like to thank the editor of the series, Janusz Kacprzyk, as well as Thomas Ditzinger and other Springer staff who supported the publication of these proceedings in the AISC series. We further acknowledge all those who contributed to the organization of the conference. We are particularly grateful to the members of the Organizing Committee for their dedication, support, and the time that they devoted to making this event a success.

Finally, we thank all the paper authors for submitting their high-quality work, and we wish to express our appreciation of conference participants for their valuable contribution to the fruitful and inspiring discussions during the conference sessions. We are delighted that participants took full advantage of the opportunity to interact, network, and connect with members of the community. We hope that this proceedings volume captures the spirit and key information of the conference and will thus be useful for researchers working in the field of man–machine interactions, accelerating new scientific achievements and discoveries driving forward further advances in our field.

October 2019

Aleksandra Gruca
Tadeusz Czachórski
Sebastian Deorowicz
Katarzyna Harężlak
Agnieszka Piotrowska

Organization

ICMMI 2019 is organized by the Institute of Informatics, Silesian University of Technology, Poland.

Co-organized with:

Institute of Theoretical and Applied Informatics, Polish Academy of Sciences, Gliwice, Poland

Honorary Patronage

Adam Czornik — Dean of Faculty of ACECS, Silesian University of Technology, Poland

Conference Chair

Sebastian Deorowicz — Silesian University of Technology, Poland

Programme Committee

Sergey Ablameyko	Belarus State University, Belarus
Ajith Abraham	Machine Intelligence Research Labs, USA
Antonis Argyros	University of Crete, Institute of Computer Science - FORTH, Greece
Hammad Afzal	National University of Sciences and Technology, Pakistan
Gennady Agre	Institute of Information and Communication Technologies, Bulgaria
Francisco Martinez Alvarez	Pablo de Olavide University of Seville, Spain
Haider Banka	Indian School of Mines, Dhanbad, India
Christoph Beierle	University of Hagen, Germany

Petr Berka	University of Economics in Prague, Czech Republic
Stefano Cerri	University of Montpellier, France
Abbas Cheddad	Blekinge Institute of Technology, Sweden
Wujun Che	Chinese Academy of Sciences, China
Davide Ciucci	University of Milano-Bicocca, Italy
Gualberto Asencio Cortes	Pablo de Olavide University of Seville, Spain
Krzysztof Cyran	Silesian University of Technology, Poland
Tadeusz Czachórski	Institute of Theoretical and Applied Informatics, Poland
Guy De Tre	Ghent University, Belgium
Kamil Dimililer	Near East University, Cyprus
Danail Dochev	Institute of Information Technologies, BAS, Bulgaria
Antonio Dourado	University of Coimbra, Portugal
Tolga Ensari	Istanbul University, Turkey
José M. Granado-Criado	University of Extremadura, Spain
Krzysztof Grochla	Institute of Theoretical and Applied Informatics, Poland
Aleksandra Gruca	Silesian University of Technology, Poland
Jerzy W. Grzymala-Busse	University of Kansas, USA
Concettina Guerra	Georgia Institute of Technology, USA
Alberto Guillen	University of Granada, Spain
Wang Guoyin	Chinese Academy of Sciences, China
Anindya Halder	North-Eastern Hill University, Shillong, India
Katarzyna Harężlak	Silesian University of Technology, Poland
Gerhard Heyer	University of Leipzig, Germany
Tinkam Ho	IBM Watson Research, USA
Chunlei Huo	Chinese Academy of Sciences, China
Alfred Inselberg	Tel Aviv University, Israel
Pedro Isaias	Universidade Aberta (Portuguese Open University), Portugal
Xiaoyi Jiang	University of Münster, Germany
Martin Kalina	Slovak University of Technology in Bratislava, Slovakia
Reinhard Klette	Auckland University of Technology, New Zealand
Petia Koprinkova-Hristova	Bulgarian Academy of Science, Bulgaria
Józef Korbicz	University of Zielona Góra, Poland
Stanisław Kozielski	Silesian University of Technology, Poland
Jacek Łęski	Silesian University of Technology, Poland
Haim Levkowitz	University of Massachusetts Lowell, USA
Evsey Morozov	Karelian Research Centre, Russian Academy of Sciences, Russia
Jakub Nalepa	Silesian University of Technology, Poland

Additional Reviewers

Rafał Augustyn	Silesian University of Technology, Poland
Michał Kozielski	Silesian University of Technology, Poland
Dariusz Mrozek	Silesian University of Technology, Poland
Marcin Pacholczyk	Silesian University of Technology, Poland

Organizing Committee

Aleksandra Gruca (Chair)	Silesian University of Technology, Poland
Agnieszka Danek	Silesian University of Technology, Poland
Paweł Foszner	Silesian University of Technology, Poland
Adam Gudyś	Silesian University of Technology, Poland
Katarzyna Harężlak	Silesian University of Technology, Poland
Agnieszka Piotrowska	Silesian University of Technology, Poland

Contents

Pattern Recognition

Bio-Data and Bio-Signal Analysis

Algorithms and Optimization

Human-Computer Interfaces

Head-Based Text Entry Methods for Motor-Impaired People

Krzysztof Dobosz$^{(\boxtimes)}$, Dominik Popanda, and Adrian Sawko

Institute of Informatics, Silesian University of Technology,
Akademicka 16, Gliwice, Poland
krzysztof.dobosz@polsl.pl, {domipop540,adrisaw877}@student.polsl.pl

Abstract. This article describes methods for text input using head mounted devices. The efficiency of six different virtual keyboards in a 3D environment were evaluated during experiments. Some of used techniques were already known earlier (dwell time, multimodality, swipes), while others are new (control of virtual cursor arrows, 8-button layout, half-star layout). Although the obtained results are not groundbreaking, the obtained results allow a good prognosis for head-based methods using VR goggles.

Keywords: Text entry · Motor impaired · Head set · Virtual reality · Virtual keyboard

1 Introduction

Text entry is a common activity of many ICT devices. Although it is easy for most people to enter text, sometimes it is a big challenge for people with disabilities. It is very important to offer different methods of introducing text to support the integration of people with disabilities in society. There are many different physical impairments that cause many obstacles in using devices intended for most people. Physical damage is a consequence of injuries and diseases. In special cases, they can even lead to quadriplegia. Therefore, to provide access to computer technology, many techniques and methods of interaction are used, among others those that facilitate the text input.

This article discusses the issue of text-entry methods based on head mounted devices (HMDs) for virtual reality. New approaches are compared to existing text input methods.

2 Text-Entry Methods for Motor-Impairment

Every text input method uses some kind of selection technique to determine the character to be chosen [13]. The main selection techniques include: direct selection, scanning, pointing and gestures.

© Springer Nature Switzerland AG 2020
A. Gruca et al. (Eds.): ICMMI 2019, AISC 1061, pp. 3–11, 2020.
https://doi.org/10.1007/978-3-030-31964-9_1

Direct selection is a technique, which enables the user to select directly one key out of whole set of keys. Unfortunately, the set of keys available for motor-impaired people is usually strongly limited. For this reason among motor-impaired people are popular ambiguous keyboards, as MultiTap [12] or T9 [8]. In these keyboards, the set of characters is divided into several groups and each group is assigned to one key. Also chording keyboards work on reduced number of keys. They require the ability to press several keys at the same time, what can be problematic for motor-impaired people. Different combinations of pressed keys correspond to different characters [9]. In the case when only a very low number (one or two) of keys are available, scanning can be used as a text entry method. This approach refers to a selection technique in which a number of items are highlighted sequentially until the desired item is selected, and the corresponding command is executed. Finally, some text entry methods use a pointing device. Many people with mobility impairments can not control the mouse directly, but they still can emulate it using joysticks, head tracking, gaze, etc. The most common solutions include: dwell time, gestural input, and multimodal interaction.

Dwell time is a hands-free input method, by which, users select a character by dwelling over it for a period of time. This technique relates to the fact that selection is always triggered when the cursor is not moving - problem is known as "Midas touch" [5]. Dwell time approach if often combined with head gestural input. It seems to be a natural way of text entry. The operation principle of head operated interfaces can base on the detection and tracking of the user's face or facial features. Modern solutions are based on computer vision and pattern recognition approaches. Captured head (or face features) movements are translated into the motion of a pointer. In [15] facial movements has been classified in two groups: rigid (e.g. rotation) and non-rigid (opening or closing of the mouth, eye winks, etc.) motions. The simplest solution can use head movements in horizontal plane [11]. In a single-row keyboard is no need to use an additional signal to confirm the selected character. It is enough to change the direction of the head movement or even just stop it.

Other approaches use a HMDs and mostly belong to multimodal interfaces. Some among existing methods can be indicated an interaction technique that enables freehand ten-finger typing in the air based on 3D hand tracking data [16]. Another solution uses special wireless data gloves for signal acquisition [1]. In the immersive virtual environment a standard QWERTY keyboard can be simulated and use with a special Pinch Gloves [4]. The text entry with hands in the virtual environment is also available using a motion controller as Leap Motion [6] that detects human gestures such as fingertip movements. Text entry techniques based on HMD can also use a game controller for scanning and selection of characters. By rotating the two joysticks of the game controller, the user can enter a text selecting characters on a circular keyboard layout in the shape of pizza [18]. Finally, HMD can be also used as support for typing that use traditional keyboards [3].

Despite many studies, text entry in virtual reality still remains a problem to be addressed. In [14] was found that though typing in VR is viable, it is constrained. Users perform better when the entire keyboard is within-view. Motion in the user's field of view negatively impacts performance. Though dwell time based selection feels natural and easy, click is the most preferred end effective way of interaction.

In the paper [17], the feasibility of head-based text entry for HMDs was investigated. In that approach, the user controls a pointer on a virtual keyboard using head rotation. Three techniques were verified: taps (pointing and tapping physical button, 15.58 WPM), dwells (pointing and dwelling over for a time period, 10.59 WPM), and gestures (using swipe-like gestures, 19.04 WPM). After long training they reached even 24.73 WPM using gesture type method. Obtained results show that head-based text entry is possible with HMDs.

3 Head-Based Text Entry Methods

Our project also focuses on controlling a pointer on a virtual keyboard, but we verified different kinds of keyboard layouts and gestures, looking for more efficient variants of existing methods.

Following methods were analyzed:

– *Dwell Time Method* (M1) - this approach uses standard layout (Fig. 1(a)). Selection of a key is realized by location the cursor point on the key area. The main purpose of this method is to allow text to be entered by swipe-like gestures cutting through buttons with the corresponding characters. The typing process takes place in two stages. First, the user moves his head towards the proper keys, and then returns the cursor to the neutral zone located in the center place of the space. While the point encounters a single button, the duration of his stay is measured, which along with the assigned letter is remembered until the end of the behavior process. After returning to the neutral zone, all saved characters are collected and passed on as one word.
– *Physical Button Method* (M2) - this approach also uses standard QWERTY layout (Fig. 1(b)), but selection is realized by pressing the additional button. First, the user moving the head must place the cursor point on an appropriate character, and then press the button. It can be a touch panel being an integral element of HMD or a trigger button in external controller.
– *Cursor Control Method* (M3) - this approach uses standard layout (Fig. 2(a)), but selection is realized by additional virtual buttons representing cursor arrows. The main control element in this method is the additional part of the standard keyboard represented by five buttons: four directional and one for confirmation of selection. The choice of one of these keys is to point it to the cursor key for the appropriate dwell time. The indication of one of the directions results in shifting the highlight of the alphanumeric key to the closest neighbor in the chosen direction. By default, the character in the center of the keyboard is highlighted at the beginning. The fifth button is responsible for confirming the currently highlighted selected character.

(a) Dwell time method   (b) Physical button method

Fig. 1. Standard layout keyboards

– *Swipe Control Method* (M4) - this approach uses non-standard layout based on an oval shape (Fig. 2(b)). The concept for this method is inspired by the Swipe approach, in which the user enters the characters in one continuous motion, marking subsequent signs along the way. However, this method differs by placing alphanumeric keys at the same distance from the eyes of the person examining.

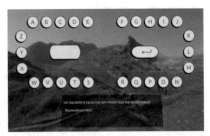

(a) Cursor control   (b) Swipe control

Fig. 2. Other keyboards

– *Half-Star Layout Method* (M5) - this approach combines the layout from direct selection with the selection method known from ambiguous keyboards. First, the user identifies the branch of the star in which the desired character is located. Then, he must move the cursor from a neutral zone located in the middle of the layout towards the mark as many times as the distance of the character in the group. The cursor does not have to reach the desired key, just move it a little. For example, selecting the letter "D" would require four moves from the center towards it. The selected character is confirmed after the specified dwell time in the neutral zone. In order to facilitate the whole process, the virtual keyboard highlights the currently selected character.

– *8-Buttons Method* (M6) - this approach uses non-standard layout being an analogy to multi-meaning keyboards characteristic for old mobile phones (Fig. 3(b)). It is composed by 8 big-size keys in the shape of ellipses. The content of these elements is the same as in the old 12-button phones. The selection of letters is realized by proper number of swipe gestures over the key area. For example, to enter the letter 'C', we have to pass the virtual cursor three times over an ellipse. In a case when the next letter comes from the same ellipse, the user has to wait until the previous letter is displayed in the text being typed. The waiting time for confirmation during the evaluation was set to one second. The vertical position of the ellipses is not accidental. The movement of head in the horizontal axis is easier and its range is larger than in the vertical axis. Horizontal swipes are more convenient.

(a) Half-star layout

(b) 8-Buttons layout

Fig. 3. Ambiguous keyboards

The Table 1 presents a comparison of the clue aspects of investigated methods: standard layout, direct selection, swipes, and multimodality.

Table 1. Genneral comparison of methods

Method name	Standard layout	Direct selection	Swipes	Multimodality
Dwell time (M1)	yes	yes	–	–
Physical button (M2)	yes	yes	–	yes
Cursor control(M3)	yes	–	–	–
Swipe control (M4)	–	yes	yes	–
Half-star layout (M5)	–	–	yes	–
8-Buttons (M6)	–	–	yes	–

4   Evaluation

4.1   Participants and Procedure

A group of 9 people participated in the study: 3 women and 6 men - in the age range 18–49. They performed multiple experiments on several selected character sets. The first one was a pangram: *the five boxing wizards jump quickly*. It makes possible to check all 26 characters of the keyboard. The remaining tests use data containing 500 pieces of text with an average length of 29 characters, and it reflects in 95% the body of English, including the frequency of characters and words [10]. Five random sentences were selected from this collection for the study:

– *valid until the end of the year,*
– *the chamber makes important decisions,*
– *did you see that spectacular explosion,*
– *jumping right out of the water,*
– *the biggest hamburger i have ever seen.*

Implemented text input methods were verified in two research environments: Samsung Gear VR (model with integrated pad that was used as an additional button in the M2 method) with a Samsung Galaxy S7 smartphone, and Setty VR 3D with a LG Nexus x5 smartphone. The time of experiments was automatically saved in the research tools.

4.2   Results

The Table 2 contains results from all tests. It takes into account the time (measured in seconds) of entering all the sentences (a total of 210 characters including spaces) and the total number of errors. The test of each method was preceded by a short training session of a few minutes. Then, sequentially, each person memorized one test sentence and made an attempt to enter it with the current method. The table contains summary times for entering all sentences.

Table 2. Result of experiments

person	M1 time	err	M2 time	err	M3 time	err	M4 time	err	M5 time	err	M6 time	err
1	432	3	387	1	845	0	546	14	387	5	1330	26
2	321	0	298	1	723	1	498	9	301	6	1123	27
3	221	2	187	0	612	0	453	8	212	7	980	12
4	267	2	243	0	745	0	523	17	254	8	880	18
5	235	1	198	0	712	1	476	4	233	6	1045	19
6	275	1	243	2	702	0	501	6	265	4	1101	6
7	301	1	259	1	704	0	554	8	256	4	1065	10
8	265	0	243	2	636	0	443	9	234	2	992	12
9	215	1	198	0	523	0	409	3	195	0	751	7

The Words Per Minute (WPM) metric is the most frequently used empirical measure of text entry performance. This metric measures the time it takes to produce certain number of words. WPM does not consider the number of keystrokes nor the gestures made during the text entry but only the length of the transcribed text [2]. Figure 4 presents WPMs calculated for each of the text entry methods.

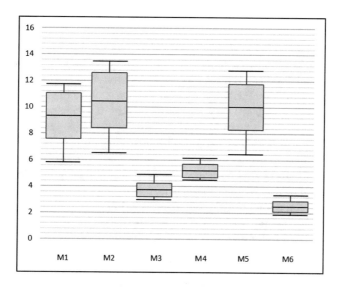

Fig. 4. Efficiency of the methods calculated in WPM

The best results were achieved for the methods M1 (9.33 WPM), M2 (10.53 WPM) and M5 (10.04 WPM). In the M1 method, the dwell time has a big influence on the result. During the experiments, it was selected after a short training for each participant and set to 0.8 second. The method with the use of an additional button (M2), as expected, worked very well. Manual confirmation of the indicated character is an effective solution, however, it must be remembered that this requires additional human interaction, which may be unavailable to people with hands disorder. The M3 method was not the best in comparison to previous approaches, but a low efficiency (3.72 WPM) should not depreciate it. M3 requires less head movements than others methods, what makes it suitable for people with a short range of head movements. The M4 (swipe control) method is relatively slow (5.20 WPM).

A high number of errors is caused by accidental typing of characters when moving the pointer over the keyboard layout. Many characters were also duplicated because after entering it, users reflexively returned through it. In the case of the M5 method based on the layout of the half-star, the relatively good result is the effect of the distribution of virtual buttons based on the statistics of the occurrence of letters in English. The highest frequency characters are placed closest to the center. The last of the verified approaches - the 8-button method, unfortunately, proved to be a missed idea (average efficiency 2.50 WPM). Despite the initial training, the participants of the study had difficulties entering the text. They accidentally entered characters while passing between them. Additionally, most participants of the research have already forgotten the layout of old 12-button keyboards and have spent a lot of time looking for characters.

5   Conclusions

All researched methods fulfill their task - they enable entering text into a computer in 3D environment. Although the obtained efficiency is worse than those obtained in other studies [17], we can suppose after longer training the results would certainly improve. Participants of the research were beginners and did not have too much time to learn new keyboard layouts and control methods before experiments. Six different techniques tested during the same sessions also did not positively affect the obtained results. Other studies shown [7], the probability of accepting a new interface is associated with how quickly familiarity can be achieved. In the case of keyboard layouts, new, sometimes more efficient designs are rarely accepted due to the difficulty of relearning.

Apart from the above-mentioned remarks, obtained results of the study are good material for further work. Next experiments can be provided by modifying the layouts of tested keyboards, the arrangement of letters, distances between virtual buttons of keyboard, respectively setting the dwell time, etc.

Acknowledgement. Publication financed by the Institute of Informatics at Silesian University of Technology, statutory research no. 02/020/BK_18/0128.

References

1. Amma, C., Georgi, M., Schultz, T.: Airwriting: hands-free mobile text input by spotting and continuous recognition of 3D-space handwriting with inertial sensors. In: 2012 16th International Symposium on Wearable Computers, IEEE, pp. 52–59 (2012)
2. Arif, A.S., Stuerzlinger, W.: Analysis of text entry performance metrics. In: 2009 IEEE Toronto International Conference Science and Technology for Humanity (TIC-STH), IEEE, pp. 100–105 (2009)
3. Bovet, S., Kehoe, A., Crowley, K., Curran, N., Gutierrez, M., Meisser, M., Sullivan, D.O., Rouvinez, T.: Using traditional keyboards in VR: steamVR developer kit and pilot game user study. In: 2018 IEEE Games, Entertainment, Media Conference (GEM), IEEE, pp. 1–9 (2018)
4. Bowman, D.A., Ly, V.Q., Campbell, J.M.: Pinch keyboard: natural text input for immersive virtual environments (2001)
5. Jakob, R.: The use of eye movements in human-computer interaction techniques: what you look at is what you get. Readings in intelligent user interfaces, pp. 65–83 (1998)
6. Jimenez, J.G., Schulze, J.P.: Continuous-motion text input in virtual reality. Electron. Imaging **2018**(3), 450–451 (2018)

7. Jokinen, J.P., Sarcar, S., Oulasvirta, A., Silpasuwanchai, C., Wang, Z., Ren, X.: Modelling learning of new keyboard layouts. In: Proceedings of the 2017 CHI Conference on Human Factors in Computing Systems, ACM, pp. 4203–4215 (2017)
8. King, M.T., Grover, D.L., Kushler, C.A., Grunbock, C.A.: Reduced keyboard disambiguating system (2001). US Patent 6,307,549
9. Lyons, K., Starner, T., Plaisted, D., Fusia, J., Lyons, A., Drew, A., Looney, E.: Twiddler typing: one-handed chording text entry for mobile phones. In: Proceedings of the SIGCHI Conference on Human Factors in Computing Systems, ACM, pp. 671–678 (2004)
10. MacKenzie, I.S., Soukoreff, R.W.: Phrase sets for evaluating text entry techniques. In: CHI 2003 Extended Abstracts on Human factors in computing systems, ACM, pp. 754–755 (2003)
11. Nowosielski, A.: Swipe-like text entry by head movements and a single row keyboard. In: International Conference on Image Processing and Communications, pp. 136–143. Springer, Cham (2016)
12. Pavlovych, A., Stuerzlinger, W.: Less-Tap: a fast and easy-to-learn text input technique for phones. In: Graphics Interface, Citeseer, vol. 2003, pp. 97–104 (2003)
13. Polacek, O., Sporka, A.J., Slavik, P.: Text input for motor-impaired people. Univ. Access Inf. Soc. **16**(1), 51–72 (2017)
14. Rajanna, V., Hansen, J.P.: Gaze typing in virtual reality: impact of keyboard design, selection method, and motion. In: Proceedings of the 2018 ACM Symposium on Eye Tracking Research & Applications, ACM, p. 15 (2018)
15. Tu, J., Huang, T., Tao, H.: Face as mouse through visual face tracking. In: The 2nd Canadian Conference on Computer and Robot Vision (CRV 2005), IEEE, pp. 339–346 (2005)
16. Yi, X., Yu, C., Zhang, M., Gao, S., Sun, K., Shi, Y.: ATK: enabling ten-finger freehand typing in air based on 3D hand tracking data. In: Proceedings of the 28th Annual ACM Symposium on User Interface Software & Technology, ACM, pp. 539–548 (2015)
17. Yu, C., Gu, Y., Yang, Z., Yi, X., Luo, H., Shi, Y.: Tap, dwell or gesture?: Exploring head-based text entry techniques for HMDs. In: Proceedings of the 2017 CHI Conference on Human Factors in Computing Systems, ACM, pp. 4479–4488 (2017)
18. Yu, D., Fan, K., Zhang, H., Monteiro, D., Xu, W., Liang, H.N.: PizzaText: text entry for virtual reality systems using dual thumbsticks. IEEE Trans. Visual Comput. Graphics **24**(11), 2927–2935 (2018)

VEEP—The System for Motion Tracking in Virtual Reality

Przemysław Kowalski[1]([⊠]), Krzysztof Skabek[2], and Jan Mrzygłód[1]

[1] KiperTech Consulting, Szwedzka 52/8, Cracow, Poland
{przemyslaw.kowalski,jan.mrzyglod}@kipertech.com
[2] Institute of Computer Science, Cracow University of Technology,
Warszawska 24, 31-155 Cracow, Poland
kskabek@pk.edu.pl

Abstract. The article presents basic assumptions of the VEEP (Virtual Entertainment Enhanced Platform) system which combines virtual reality goggles with motion tracking system based on several of RGB cameras. We present basic algorithms for the system calibration and maintenance. The initial performance results were also given.

Keywords: Computer vision · Motion capture · 3D localization

1 Introduction

Virtual reality (VR) systems make it possible to immerse in the realistic 3D world. The technology uses just a computer screen or more advanced modern devices such as Head Mounted Displays (HMD). The Oculus goggles can be very useful for this purpose as they are wireless and provide pretty good resolution of displays. Another challenge for virtual game systems is a user interface for tracking the player movements. Our aim was to develop the tracking interface which would increase immersiveness and give intuitive and realistic feedback to the system.

In the bibliography we can find many different ways of feedback for motion tracking in VR. One of the popular solutions is Microsoft Kinect [13] which uses the depth images to obtain the pose estimation. Such installation can be extended to Multiple Depth Camera Approach (MDCA) [14] to track human motion from multiple Kinect sensors that can significantly improve the tracking quality and reduce ambiguities caused by occlusions. Another way to track the body motion are multiple color-filter aperture (MCA) cameras [6] where depth information as well as color and intensity is gathered in the single-camera framework. We also noticed the usage of InfraRed (IR) cameras for this purpose [8]. Another solution for movement tracking is GPS localization useful only in outdoor environments and also more universal Wi-Fi or bluetooth transmission. In some cases we also found the camera tracking which relied on the routines from the OpenCV library [4].

© Springer Nature Switzerland AG 2020
A. Gruca et al. (Eds.): ICMMI 2019, AISC 1061, pp. 12–22, 2020.
https://doi.org/10.1007/978-3-030-31964-9_2

The alternative to the above solutions can be low cost human motion capture system which uses a set of distinguishable markers placed on several body landmarks and the scene is captured by a number of calibrated and synchronized cameras [2]. In our approach we also utilize similar approach to localize a human wearing VR goggles. Another similar approach is commercially available from Vicon [15], although our system should be cheaper.

2 The System Outline

The Virtual Entertainment Enhanced Platform (VEEP) combines virtual reality (VR) visualization with motion tracking system developed in our company. Compared to the many works on human motion [5], centered on a precise joint movement, in our application the key was to achieve a low latency with limited financial expenses. Spatial sounds and additional effects, such as humidifier or blower are also planned – all this to make user immersion even deeper.

The intention of the project was to provide the system for motion tracking at low level costs to make it popular and widely used with VR devices. The remaining two parts of the project: VR goggles (eg. Oculus GO[1]) and game engine software (eg. Unity 3D[2]) are commercially available. One of the important project's challenge is the possibility of integrating existing technologies to obtain technology with new possibilities.

2.1 VR Goggles

In our project we focused on cheap, standalone VR goggles. The popular VR tethered goggle were not acceptable due to dynamic movements of players in the scene. In fact we had two options: standalone or mobile[3] headset [9]. First group is represented by Oculus GO and Lenovo Mirage Solo With Daydream, and the second group is based on mobile phones such as Samsung Gear VR working with Samsung Galaxy smartphone.

We tested both Oculus GO and Samsung Gear VR and the tests suggested that the higher quality of visualization would be obtained using Oculus GO, even for the comparison to the newer models of Samsung Galaxy. The Samsung Gear quality was strongly dependent on the used mobile phone. We do not exclude the use of other platforms in the future—because we try to create software in a generic way.

2.2 Game Engine

Looking through the available game engines we focused on the same aspects: easy for game development, good performance and high image quality. Another

---

[1] https://www.oculus.com/go/.
[2] https://unity3d.com/unity.
[3] The headset uses mobile phone.

advantage is availability for multiple platforms and especially for Linux and Android.

We chose between two environments, which are popular and available for many platforms: Unreal [3] and Unity [1]. Both are ready for VR goggles. Our tests suggested that Unreal gives better performance (image quality and speed (speed—see also [11])), but on the other hand, Unity reduces time and costs of preparing the initial software version. Moreover, Unity platform is free even for small commercial use and open source with many libraries and variety of drivers for external devices—for these reasons we decided to use Unity.

2.3 Motion Tracking

To get the impression of VR immersion, we should achieve fluent motion capture of players with at least 60 fps acquisition [7]. The expected frame rate for cameras in our system should be no less than 60 fps with the smallest possible latency.

There are two popular interfaces that provide satisfactory transfer of uncompressed images: GigE Vision and USB 3.x. We decided to use USB 3.x as it gives the better transfer and is more popular and available for many platforms.

The assumption for our system is that it uses many cameras which are not predetermined, on the one hand, to choose economic solutions (parameters of the cameras vs price) on the other to maximized the observed area from at least 2 cameras. We also chose active markers working in visible lighting (Fig. 1). For each player we are using 5 markers, now—for body, hands, and legs (Fig. 4). The set of markers for each player may be redefined. Markers have six distinguishable colors: pink, green, blue, orange, cyan, light green/light yellow.

Fig. 1. The markers for VEEP project.

Our system consists of four applications (see Fig. 2):

- *Calibrator*—used to calibrate cameras—the output is the set of calibration parameters (Tsai camera model) stored in files.
- *MarkersFinder*—used to find markers in the camera images. One instance of MarkersFinder is used for one camera. The output is the set of 2D markers.

- *Coupler*—calculates 3D positions from 2D markers using camera parameters. The output consists of 3D positions assigned to players and players' body parts.
- *ViMobile*—visualization application (built using Unity 3D).

In addition we prepares some auxiliary applications for tests and system verification.

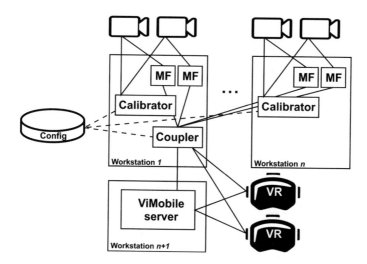

Fig. 2. The scheme of the VEEP system. For each workstation (WS) with cameras we are using MarkersFinder and Calibrators. For the system we use one Coupler and one ViMobile server. Config. means configuration data—set of files accessible for Coupler and MarkersFinders.

3   Localization of 2D Markers in Images

We proposed the algorithm for 2D marker localization taking into account the requirements of high computation time and quick configuration. To achieve these requirements we use active markers in a darkened room. Marker patterns are stored in separate configuration files.

Our algorithm consists of three stages. In the first stage we check chosen pixels, if there can be characteristic points of the markers. Finding such point we assume, the pixel may be a part of the marker. To check it we compare pixel with the given pattern.

In the second stage, starting from the "characteristic point" we fill the marker surface. The algorithm uses three criteria: minimal and maximal values of the component of the pixels, threshold for pixel values change, and the threshold for minimal light value. The importance of the criteria depends on the markers pattern definition. Filling the surface, we calculate: the center of the marker, its surface, and the average color.

In the third stage we filter results, analyzing the size of the markers (extreme size suggests error) and mutual locations (very close markers are probably the same marker, partially occluded).

Such algorithm has the following advantages:

- The first step is very quick and most of the pixels in the input image are rejected from the further analysis.
- In both initial stages we use pattern defined by users—the algorithm is flexible.
- The second stage is flood fill—it gives center of the marker with *subpixel* accuracy as we use float numbers.
- In the second stage we calculate color of the marker, as a mean normalized RGB (the components R, G, B are divided by their sum—pixel brightness)— it gives better accuracy than determination of the color using one pixel value.
- The recognizable colors of the markers are defined as a set of files with markers patterns and not fixed in the algorithm.

The output of the algorithm is the list of markers in the form of a stream of compressed binary metadata (to take the least possible data transfer medium). Each marker is described by: color, coordinates x, y (as float numbers), camera number, markers size, frame number, additional test information, e.g. time stamps.

The calculated markers are sent via UDP protocol to the *Coupler*.

4   3D Position Calculation

The 3D position calculation is based on the Tsai camera model [12]. The input data is a list of n corresponding pairs of points in the scene (x_{wi}, y_{wi}, z_{wi}) and also the corresponding sensor coordinates (X_{fi}, Y_{fi}). The algorithm consists of two main stages: (1) computing 3D orientation $(r_1 \ldots r_9)$ and position (T_x, T_y) as a sequence of partial minimizations, (2) computing effective focal length f, distortion coefficients κ_1, and position T_z. The second step is realized as a non-linear least squares fitting problem. Calibration calculations are performed separately for each pair of cameras.

We assume that the number of markers in the scene is given and we know how many users, with the defined markers will appear on the scene. The information is used to remove unnecessary markers, and to match the found markers with the relative position in the scene. For example, we know that there are two green markers, one for right hand, and one for left leg of the player number 1, so we match the markers best fit in such starting positions. We assume that usually the hand is higher than the leg at the beginning of the system operation.

4.1   Calibration and Tsai Model

Calibration procedure is led by operator. We start with automatic Tsai calibration, but the error in feature points positions usually lead to the set useless parameters which must be adjusted, so the manual calibration is performed, where the operator sets basic Tsai parameters (three for angle, three for translation), matching the pattern to the image (see Fig. 3).

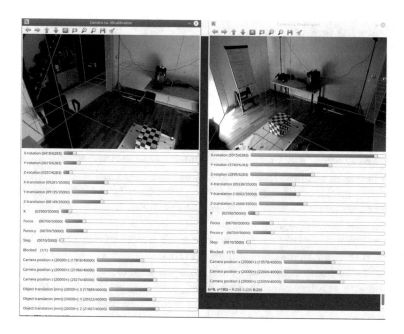

Fig. 3. The calibration process for two cameras.

4.2 Calculation of 3D Points

We match all the markers of the same color. For each pair of markers (from different cameras) we calculate the minimal distance in 3D between the relative positions of markers. For real markers and proper calibration parameters, the distance should be very small. We defined such max value δ for the distance error to decide if the matching is appropriate.

To calculate 3D position we use two 3D lines, determined by camera center and a point (from 2D marker position) in the image plane. Both lines are represented as vectors: $l_1 = \mathbf{a} + \mathbf{b}s$ and $l_2 = \mathbf{c} + \mathbf{d}t$. We are looking for minimal distance between both lines assuming the calibration and calculation errors makes the line non-intersecting. We can calculate s and t:

$$s = -\frac{((\mathbf{b} \cdot \mathbf{d})(\mathbf{b} \cdot (\mathbf{c} - \mathbf{a})))/(\mathbf{b} \cdot \mathbf{b}) - \mathbf{d} \cdot (\mathbf{c} - \mathbf{a})}{(\mathbf{b} \cdot \mathbf{d})^2/(\mathbf{b} \cdot \mathbf{b}) - \mathbf{d} \cdot \mathbf{d}} \tag{1}$$

$$t = \frac{\mathbf{b} \cdot (\mathbf{c} - \mathbf{a}) + \mathbf{b} \cdot \mathbf{d}s}{\mathbf{b} \cdot \mathbf{b}} \tag{2}$$

Using s and t we can calculate the closest points on both lines and the distance between them (δ). The distances are used to determine if the 3D point position is real position of the marker, or it is a false match.

If the 3D points were found, we go to the furthers steps of analysis. We try to match the same points calculated using another pair of cameras. For example, from 3 cameras and two real markers in the same color, we should obtain 6 correct 3D positions, corresponding to two real 3D points. In fact the number of correct positions could be bigger as some wrong matches are theoretically acceptable; or smaller if markers are occluded. The algorithm checks the groups of 3D points calculated using coherent collection of 2D markers.

4.3   Tracking the Markers

Markers are tracked in 3D using two features: color and the Euclidean distance between new and old positions. Then the nearest new marker is matched to the old marker position, with limited jump value which is the maximum distance between old and new position. If marker was occluded in many frames, the jump limitation can be skipped.

In our motion capture system we use stick model of the human body (see comparison of the human body models in [10]).

Markers are further grouped. Each group represents one person (Fig. 4). In our convention the left hand and right leg, and—respectively—right hand and left leg, have the same colors, so it can be incorrectly matched. For each group we check if the positions of markers are in proper relation, and if not—we change the positions.

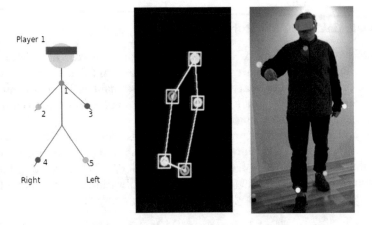

Fig. 4. The set of markers used for the player—from the left: scheme, the real scene observed in the system, and a real player.

5 Experimental Environment

We prepared experimental environment consisting of two workstations, two VR goggles (Oculus GO), and four cameras.

The workstations use Linux Mint operation system. The first configuration: Intel(R) Core(TM) i5-3470 CPU, 3.2 GHz and 8 GB RAM and the second: Intel(R) Core(TM) i7-3770 CPU, 3.4 GHz and 32 GB RAM. Two cameras are connected to each workstation. For each camera we run *MarkersFinder* module. In one of the computers we run *MotionCapture*, *MarkersFinder* and *MotionCapture*. The modules are connected via UDP protocol. The configuration parameters are stored in files—the same for both computers. In one computer *ViMobile* module is also running as a server of 3D actions.

We use two Oculus GO goggle with ViMobile applications and four cameras PS4 Eye—only one camera in such stereovision set is used. The cameras run in mode 640 × 400 pixels at 120 fps.

6 Efficiency

We assumed to determine the 3D player position about 60 fps rates. It means that the system should calculate each 3D marker position in time 16.33 ms.

6.1 Network Efficiency

The average latency between two computers (markers 2D send to Coupler) was 0.187 ms (maximum: 0.206 ms). The average latency between computer and VR goggles (markers 3D send to visualization) was 7.622 ms, maximum latency was 97.3 ms (but the median is 2.45 ms).

6.2 Tracker Efficiency

We have checked work times of one MarkersFinder (MF—one from two working concurrently). In table below times are given in ms. Tracking itself takes only about 1 ms, while image acquisition takes about 6.67 ms, cooperation with Coupler—about 2.77 ms; visualization—5.67 ms. We can use MarkersFinder without visualization and fetching data from Coupler (it is important only for tests)—that actions takes 8.22 ms.

Action	Module	Sample times [ms]									Average [ms]
		1	2	3	4	5	6	7	8	9	
Image acquisition	MF	6	7	7	4	9	6	7	7	7	6.67
Markers tracking	MF	1	1	1	2	0	1	1	1	1	1.00
Send to Coupler	MF	1	1	0	0	1	0	0	0	0	0.33
Get from Coupler	Coupler	8	0	10	0	4	0	0	0	0	2.44
Convert from 3D	MF	0	0	0	0	0	0	1	0	0	0.11
Show images	MF	1	2	1	2	1	1	1	1	1	1.22
Additional actions	MF	1	4	0	9	0	7	5	5	9	4.44
Summary		18	15	19	17	15	15	15	14	18	16.21

6.3 Coupler Efficiency

We tested the execution times for each frame with of time consumption of selected actions (see table above). The values in table were averaged to milliseconds. We can see that the *Coupler* working time is the meaningful value as well as time of image acquisition from *MarkersFinder* module. The *Coupler* time includes the calculation the 3D position and assigning it to the proper part of the player body.

7 Accuracy

The system accuracy depends mostly on calibration quality and position errors caused by noise in input images.

We checked the changes of position caused by image noise for more than 4000 frames and two fixed markers (blue and orange). The positions were calculated using 2 cameras (the table below). The distance between extreme positions of the marker was 9 mm. The error caused by noise depends on the lighting condition.

Value	Marker no	X position [cm]	Y position [cm]	Z position [cm]	Total [mm]
Minimal value	1	−436.8	115.1	326.8	
Maximal value		−436.7	115.6	327.6	
Difference		0.18	0.49	0.74	9.025
Minimal value	2	−403.5	94.6	310.0	
Maximal value		−403.4	94.9	310.4	
Difference		0.17	0.30	0.41	5.34

8 Summary

We built the simple and efficient visual motion tracking system connected with VR goggles to prepare immersive 3D environment in real place and use it for interactive VR visualization. The system is scalable with regard to the number tracking cameras, workstations and the room geometry. We obtained satisfactory frame rate. Our system is low cost—it does not require high computing power, and uses easily accessible cameras with Sony PS4 Eye. The system is still a prototype: we need to improve its accuracy and stability especially in the case of multiplayer acting.

Acknowledgements. This work was supported by the NCBiR (The National Centre for Research and Development), project VEEP (Virtual Entertainment Enhanced Platform), no. POIR.01.02.00-00-0155/17-00.

References

1. Bond, J.G.: Projektowanie gier przy użyciu środowiska Unity i języka C#, trans. Gliwice, second edition, Jacek Janusz, Helion (2018)
2. Canton-Ferrer, C., Casas, J.R., Pardàs, M.: Towards a low cost multi-camera marker based human motion capture system. In: 16th IEEE International Conference on Image Processing (ICIP) (2009)
3. Cookson, A., DowlingSoka, R., Crumpler, C.: Unreal Engine w 24 godziny. Nauka tworzenia gier, trans. Joanna Zatorska, Helion, Gliwice (2017)
4. De Schepper, T., Braem, B., Latre, S.: A virtual reality-based multiplayer game using fine-grained localization. In: GIIS IEEE (2015)
5. Kolahi, A., Hoviattalab, M., Rezaeian, T., Alizadeh, M., Bostan, M., Mokhtarzadeh, H.: Design of a marker-based human motion tracking system. Biomed. Sigal Process. Control **2**, 59–67 (2007)
6. Lee, S., Lee, J., Paik, J.: Simultaneous object tracking and depth estimation using color shifting property of a multiple color-filter aperture camera. In: International Conference on Computing, Management and Telecommunications (2013)
7. Madhav, S.: Game Programming Algorithms and Techniques: A Platform-Agnostic Approach. Addison-Wesley Professional, Boston (2013)
8. Marks, S., Estevez, J.E., Connor, A.M., Towards the holodeck: fully immersive virtual reality visualisation of scientific and engineering data. In: ICIVC ACM (2014)
9. Parisi, T.: Learning Virtual Reality. O'Reilly Media, Sebastopol (2015)
10. Green, R.: Spatially and temporally segmenting movement to recognize actions. In: Rosenhahn, B., Klette, R., Metaxas, D. (eds.) Human Motion, Understanding, Modelling, Capture, and Animation. Springer, Dordrecht (2008)
11. Smid, A.: Comparison of Unity and Unreal Engine. Bachelor Thesis, Czech Technical Univesity in Prague (2017)
12. Tsai, R.Y.: An efficient and accurate camera calibration technique for 3D machine vision. In: Proceedings of IEEE Conference on Computer Vision and Pattern Recognition, Miami Beach, FL, pp. 364–374 (1986)
13. Woodard, W., Sukittanon, S.: Interactive virtual building walkthrough using Oculus Rift and Microsoft Kinect, University of Tennessee, Martin Department of Engineering. Technical Report (2013)

14. Zhang, L., Sturm, J., Cremers, D., Lee, D.: Real-Time Human Motion Tracking Using Multiple Depth Cameras (2012)
15. Origin by Vicon. https://www.vicon.com/motion-capture/vicon-reality. Accessed 3 June 2019

Immersive Virtual Reality for Assisting in Inclusive Architectural Design

Ewa Lach[1]([✉]), Iwona Benek[2], Krzysztof Zalewski[2], Przemysław Skurowski[1],
Agata Kocur[2], Aleksandra Kotula[1], Małgorzata Macura[1], Zbigniew Pamuła[1],
Mikołaj Stankiewicz[1], and Tomasz Wyrobek[1]

[1] Institute of Informatics, The Silesian University of Technology, Gliwice, Poland
{Ewa.Lach,Przemyslaw.Skurowski}@polsl.pl,
{alekkot200,malgmac902,zbigpam558,mikosta165,tomawyr482}@student.polsl.pl
[2] Faculty of Architecture, The Silesian University of Technology, Gliwice, Poland
{Iwona.Benek,Krzysztof.Zalewski}@polsl.pl, agatkoc620@student.polsl.pl

Abstract. Immersive Virtual Reality (IVR) technology is capable of simulating highly realistic environments and affecting behavioral realism via the creation of the feeling of immersion in a virtual world. Because of that IVR has got a great potential as a tool for evaluation of architectural designs. IVR allows to navigate in the designed space and interact with its elements as if they were real. In this paper, we demonstrate that IVR can assist during the Inclusive Design process and improve the architects' understanding of the needs of different users, especially those that are older or disabled. Inclusive design promotes products and environments that are inclusive for all people. Architects require in-depth insight into how particular groups of people experience the designed environment and how they interact with it. IVR can help with that.

Keywords: Immersive virtual reality · Inclusive design · Architectural design process · Assisting architects in design process

1 Introduction

Architectural design is intensely multifaceted and challenging. Architects have to reconcile multiple objectives which are often in conflict: fulfill customer wants and needs while also adhering to the functional, aesthetic and economic requirements. Inclusive design features (i.e. features that create an environment space that is usable to the greatest number of individuals, regardless of their functional abilities) are often added as an afterthought at the end of a design process if they are considered at all [11,17]. This needs to change. Architects and designers are facing new challenges as a result of the growing number of older people and the corresponding increase in health problems in the society. According to the World Health Organization [25] about 15% of the world's population has some form of disability and between 2.2% and 3.8% of adults have significant difficulties in functioning. What's more, rates of disability are increasing, due to, among other

© Springer Nature Switzerland AG 2020
A. Gruca et al. (Eds.): ICMMI 2019, AISC 1061, pp. 23–33, 2020.
https://doi.org/10.1007/978-3-030-31964-9_3

things, population aging. According to the UN, more than 46% of older people - those aged 60 years and over - have disabilities. According to [24] the number of older people has increased substantially in recent years in most countries and regions, and that growth is projected to accelerate in the coming decades.

The obvious conclusion from the above facts is that the needs of seniors and people with disabilities should be included in architectural designs. Architects and other designers of the built environment are well positioned to be the advocates for inclusive design, to drive change to the built environment and thereby enhance social equality [9]. The problem is a lack of understanding from designers, particularly during the planning and design phases, of how their designs will be experienced by people with disabilities [2,9,17,22,23].

Full-scale physical mock-ups are created during the construction phase and are of limited use for making extensive decisions regarding the functionality of the designs [8]. Three-dimensional visualization tools offer the opportunity to comprehend proposed designs more clearly and test them more thoroughly during the planning and design phases, thus enabling the greatest possible influence on design decision making [3,4,19]. Especially Immersive Virtual Reality creates the possibility of replacing the physical mock-ups with digital ones leaving the interactive and immersive experience [3,4,8,13,15,18].

In this paper authors argue that IVR not only allows the user to experience proposed designs like in real life but can also simulate experiences from the perspective of a disabled person, augmenting and modifying the user's experience. As a result, it can enable designers to create a better, more inclusive design. As a test bed, the IVR tool is presented that allows the user to perform everyday activities in a modeled apartment. The perception of space in the application is simulated from the perspective of a person using a wheelchair.

The paper continues by presenting the problem of inclusive design in Sect. 2. Section 3 introduces the computer-aided architectural design process. Section 4 presents the IVR tool for assisting architects in designing for disabled people. Section 5 concludes the paper.

2   Inclusive Architectural Design

The world is undergoing a major process of demographic alteration involving the aging of the population [24]. Older people are a diversified group primarily in terms of health and their physical and intellectual skills. However, there are higher disability rates among older people as a result of the accumulation of health risks across a lifespan: diseases, injuries, and chronic illnesses. As human ages, vulnerability to disability also substantially increases. For instance, in old age, our vision and hearing capacity deteriorates, the risk of falling increases, and the consequences of injuries are more serious [25].

Inclusive design has got the potential to help address the challenges of a diverse and aging society in a sustainable way. Inclusive design also referred to as universal design, accessible design, and barrier-free design is an international concept that is intended to allow products and environments to be inclusive of all people. For residential design, it means attending to a dynamic range of people and abilities. Housing must be adaptable to accommodate the different needs and requirements of the users. Individual characteristics such as strength or agility should not prevent a person from safely using and enjoying all features of their home. The architectural features should be designed inclusively to equalize accessibility, privacy, security, safety and usability of those spaces.

The creation of an environment that meets both the needs and preferences of older people is a great challenge for designers. It requires a detailed and insightful approach and must be constantly adaptable to the ever-changing skills and needs of seniors. It is also necessary to take into account the common types of disabilities like motor, auditory, visual and mental ones. Nowadays, societal discourse on housing and care for older people reflects growing attention for their autonomy and individuality as well as integration with the community and participation in it. The aim is allowing people to remain independent for as long as possible [1,16]. According to the theory of environmental stress and place attachment [5], older people feel best in a familiar environment that is most well-known to them, that is: in their own apartments. Changing the place of living is always a traumatic experience for the elderly so they ought to stay in their apartment for as long as possible. For this purpose, housing solutions like aging in place [26] have been developed to promote the design and building of dwellings that will suit people from birth to death, regardless of their changing needs and levels of ability.

While designing houses suitable for the elderly, architects can refer to a series of guidelines for the inclusive design [12]. At the basic level, designing more inclusive spaces and buildings is expected to reduce the need for adaptations at a later stage. The built environments can be regarded as care-negative, -neutral or -positive. Care-negative environment occurs if additional care is needed to overcome architectural barriers, leading to people being less independent and their quality of life suffering due to the built environment (for example the place not being accessible for a person on a wheelchair). In a care-neutral environment, people do not require help to overcome the environment's physical and other shortcomings. An environment can be regarded as care-positive if its inclusive design allows for, or even encourages, user's independence.

The limited adoption of inclusive design in architecture practice [22] suggests that there are still many challenges to address, related to the perceptions of inclusive design, its practical applicability, and its infusion in architectural education. Very often designers see inclusive features as restrictions on their creativity, misaligned with the aesthetics of their design. Additionally, inclusive features require more space and money. Designers focus on preferences of their clients (the building developer/owner, who also usually do not understand principles of inclusive design) not considering that the end-users of the building would also, in the end, be their 'clients' [17]. Most architects associate inclusive design with laws, certificates, rules, standards, and regulations so it is, to them, solely mathematical calculations and measurements [11]. Without previous exposure to, or experience with, the everyday interaction of people with disabilities in the built environment, the architects generally do not consider environmental and social barriers faced by those groups in the population. Designers need to understand how diverse people are affected by the built environment because qualitative dimensions are just as important as quantitative ones. Architects require in-depth insight into how particular groups of people experience and interact with the designed environment and the building codes and legislation cannot be the main source of information about this interaction.

Higher Education institutions are ideal places for raising awareness among professionals and other society members. Inclusive Design should be taught in all places where the future professionals, whose work will shape the future built environment, are educated [9,23]. Teaching should include building empathy by inviting the people with different needs and listening to their experiences and opinions as well as promoting practical learning experience of being disabled (using wheelchairs, blindfolds, canes, etc.) [2,17].

In recent years it has become more widely recognized that the elderly and disabled population should be included in any design process affecting their quality of life [1,6,11]. Various examples show that fewer mistakes and better results are obtained if older people participate in the decision-making and design processes [14,16]. Designs should be evaluated from the perspective of people with disabilities.

3   Computer Aided Architectural Design Process

There exist advanced professional computer applications with various functions and capabilities used to support architectural design. The simplest software supports only two-dimensional designs (e.g. AutoCad LT) while more advanced options (e.g. AutoCad) also supports 3D modeling. The software of this generation operates with basic graphic elements such as lines, points, basic geometrical figures, etc. There is no direct relation between the flat drawing and a 3D model - these elements are not interconnected, i.e. the model change is not reflected automatically in the flat drawing and vice versa. These tools are basically a more advanced, digital version of a drawing board and now they are being replaced by software in which we operate three-dimensional representation and information about the object and its elements - the so-called BIM (Building Information Model) (e.g. Revit, Allplan, ArchiCAD, etc.). This is a significant improvement in terms of technological advancement, but also its usage is more complex, requires more sophistication from the user and equipment. Its essential feature is the manipulation of parametric three-dimensional objects such as doors, windows, walls, structural elements, etc. additionally enriched with detailed information on any physical, finishing, safety, etc. features. The predominant feature of BIM software is the ability to integrate all project disciplines within a single model, making it easier to coordinate the project, and in particular to verify inaccuracies and collisions. Ultimately, there is also the possibility of visualizing the model, exporting elements for CAM production directly from the model, omitting paper documentation, as well as managing the construction from the application level. Although these applications seem to be a complete tool, it is worth noting some limitations: the software is used primarily for the purpose of documentation development, so its proper application starts after the creative phase. In spite of its sophistication, it does not offer testing capabilities in terms of utility, e.g. in the field of ergonomics. For these reasons, the mentioned tools have a limited and auxiliary role in the designing phase. The character of this phase requires a conceptual approach, for which the possibility of easy modification of assumptions (layout of rooms, furniture, walls) is crucial. Software dedicated to conceptual designing should also provide intuitive, easy and fast service. In contrast to programs used to develop documentation, the choice of this type of software is limited. The most popular program is SketchUp. In addition to fairly precise modeling of three-dimensional geometry, the program allows viewing the model from any perspective, generating views and sections. However, it is not possible to generate CAD documentation directly from the program, only to export geometry to dedicated programs. A separate category are programs for visualizations. The basic, traditionally used software has got the possibility of photorealistic visualization of static scenes (Blender, V-ray) or animation (3D Max). The most advanced tools (e.g.: Lumion) enables photo-realistic visualization in real time and elements of current interaction with the model (the ability to create an ad hoc pathway - as in a computer game). The software allows you to assess the space created in terms of aesthetics and formal use with regard to environmental characteristics, such as lighting, etc. However, there is no possibility of direct interaction with the model and its elements - door opening, moving elements, etc. which significantly limits its application.

It can be noticed from the above discussion that comprehensive architectural designing with IT tools requires the application of a set of tools dedicated to various purposes. At the same time, none of the groups of programs offers tools to evaluate ergonomics and functionality, in particular ones dedicated to the elderly and the dysfunctional. Reviewing the literature indicates that there is recognition of the need for evaluation tools during the design phase and that the new technologies such as Virtual Reality (VR) are believed nowadays to be the most efficient for architecture information representations [21].

In a virtual environment, humans can communicate and transfer knowledge without time and space limitations that exist in the real world while saving time and money and minimizing health risk. We can witness a trend of using VR for construction risk identifications and safety training - 3D building visualizations allow to assess building site conditions and recognize possible hazards before proceeding with the construction [3]. VR can be widely used in the education of architecture students to allow them to experience being on a construction site, investigate the architectural details closely, and test what they have learned in an interactive and immersive environment [15]. Students can also evaluate their designs. VR allows to easily test hypothetical designs, explore new possibilities and compare design options. VR can support the designer in making many important decisions that can affect the final quality of the built environment, supporting the implementation of the best solutions that could otherwise be missed at this stage and later prevented from being implemented [4,8,19].

We can recognize the five most important features of VR that makes it very useful in the evaluation of built environments: visual realism, immersion, interactivity, low cost, and safety. Present day VR-technology is technologically capable of simulating highly realistic, highly detailed, and highly 'complex' environments. VR allows for thorough observations, accurate behavior measurements, and systematic environmental manipulations under controlled laboratory circumstances. One of the cardinal features of VR is the sense of an actual presence in, and the control over, the simulated environment. To increase the sense of immersion, traditional VR systems that typically use computer screens to display virtual environments can be replaced by IVR systems that allow users to walk through and interact with the three-dimensional data. While traditional VR systems provide high visual realism, IVR can additionally simulate behavioral realism (e.g. via reaction to a user's body and head motion by changing the virtual view). In IVR the users perceive themselves as being surrounded by the 3D world. Authors in [18] prove that using IVR gives a better spatial perception of virtual space than using traditional VR. It is also presented in [8] that a lack of immersion in a scene may result in missing some spatial shortcomings in the design by the reviewer. In this case, the VR monitor display of the hospital room doesn't allow the user to recognize the strike hazard of the shelves as opposed to IVR.

In following years we should see greater application of IVR in architecture thanks to the emergence of high-quality, affordable head-mounted displays (HMDs; e.g. Oculus Rift, HTC Vive, Samsung Gear VR) which are more accessible than the formerly-used IVR Cave Automated Virtual Environments (CAVE), where the images were projected on the walls, floor, and ceiling of the room that surrounds the viewer. HMDs use a visual display that covers the full visual field of the user, and a head tracking device which tracks the position and orientation of the user's head. In [13] author presents a tool that allows for an interior design process in an IVR environment.

In this paper, we want to focus on the inclusive design process that enables the evaluation of the designed space during the designing phase. Authors in [8,19] used VR technology to allow people with disabilities to assist with designing by evaluating if designs suit the requirements of future users. People on wheelchairs were asked to test proposed solutions. In our study, we propose a use of IVR that enables the evaluation of the built environment from the perspective of an elderly disabled person which can be performed by the designer regardless of his or her abilities. We want to demonstrate that IVR can help an architect with designing, giving him or her the opportunity to become the person with any physical handicaps needed for that assessment.

4   IVR in Architectural Design Case Study

The paper presents the results of the project titled PERSPECTIVE 60+ that was carried out by an interdisciplinary team including students and tutors of two faculties of the Silesian University of Technology in Gliwice: Faculty of Architecture and Faculty of Automatic Control, Electronics and Computer Science.

The main aim of this project was the implementation of the achievements from the field of IVR in order to assist the architect with the design of a built environment for an elderly person with disabilities using the principles of inclusive design.

We want to show that IVR technology can be used for this purpose, because it allows for the desired simulations: navigation in the designed space, interaction with elements of that space, and simulations of different perceptions of the presented reality, including the perspectives of older and disabled people. It is possible to simulate various disabilities - motor, visual, etc.

We decided to design a residence of an elderly person as it is the main place that decides the quality of life of an elderly person and as such, it can promote his or her independence or contribute to disabling situations, producing "architectural disability". As indicated in [10] the majority of accidents in the EU in 2015 took place at home, which was particularly visible in older generations.

Figure 1 shows the steps of the design process that was used by the authors. We explored the utility of IVR for evaluation of the design from the perspective of an older person with disabilities. For study, we have chosen to simulate a motor disability as it is the most common one. According to [20] those with motor impairment constitute almost half of the population of impaired people in Poland. In our project navigating in the modeled world takes place on a wheelchair.

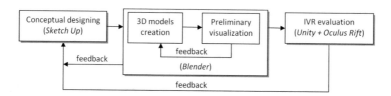

Fig. 1. Architectural Design Process.

The main objective of the architecture part of the interdisciplinary team was to design the residence for a wheelchair-bound elderly person following the guidelines for inclusive design (Fig. 2). We investigated if utilizing IVR capabilities gave us a better understanding of the design from the point of view of its intended user.

In the project, the rules of inclusive design for an impaired person were identified, included, implemented and evaluated. IVR allowed for highly-detailed observations of the evaluated design. What's more, it enabled an assessment of accessibility, navigability, and usability of the space of the apartment.

Survey of 44 housings for older people in England found that the greatest barriers to inclusion were all related to poor accessibility [16]. The main reason for that is the fact that the functional range for the elderly and disabled people is limited when compared with that of the young and abled people [7]. So it

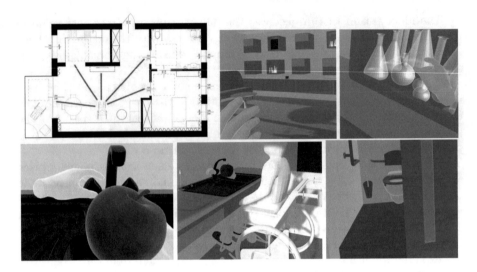

Fig. 2. Projection of the modeled apartment and UCS (1). Screenshots from the IVR application: a system of wall cabinets that can be moved on the rectangular plan (2) and assessment of their accessibility (3); using sink: 1st- (1) and 3rd- person view (3); and the sliding doors in the house (6).

is not a surprise that according to the survey conducted among elderly motor-impaired people, their caretakers and physiotherapists regarding their preferences for general design features of a kitchen the most crucial feature for them is the adjustable height of the furniture [6]. Following this information, all cabinets used in the designed home have got a height-adjusting mechanism. Using IVR we could assess the functional range of hands and thus the availability of cabinets by removing objects from them while in first-person view (Figs. 2.2 and 2.3).

Our evaluation tool also enabled us to explore the difficulties people on wheelchairs have with opening doors. Performed in IVR tests confirmed the superiority of the sliding doors over the more widespread swinging doors. Additionally, during the IVR step of the design process (Fig. 1), a significant error in the sliding doors design was detected which was not detected in the previous two steps of the design process: the person in the wheelchair was not able to go through the passage, despite it being ostensibly designed with a wheelchair in mind. The width of the doorframe was correct for a swinging door or a sliding door which can disappear into the wall completely - the error resulted from the placement of the door handle, which caused the door to be unable to slide into the wall completely, and thus blocking part of the doorframe (Fig. 2.6). The mistake stemmed from exchanging swinging doors for sliding doors during the design process and overlooking how they would limit the functional doorframe width, showing how lack of immersive visualization and thorough simulation can introduce fundamental failures even into designs which ostensibly focus on people with disabilities from their very inception.

Accessibility is closely related to the problem of sufficient navigation space that is especially important to people using wheelchairs. With IVR we had the ability to move around the designed space, supplemented by simulations of collisions with its elements which enabled the assessment of space availability and the visual assessment of the overall impression: passage widths, passage geometry, room dimensions. In our test apartment, we investigated if there is enough space for motion and assessed the quality of that motion. We removed thresholds, carpets and other floor-related obstacles that were hindering a movement of a wheelchair. We studied if there was enough space under the table, kitchen counter and sinks (Figs. 2.4 and 2.5) for a wheelchair and wheelchair's user legs. We analyzed the availability of wardrobes in the entrance area. We examined if the so-called user control center (UCS) allows for observation of the apartment's interior and its surroundings and access to everyday objects (Fig. 2.1).

The IVR application used in this project is intuitive to use. Oculus Rift HMD visualizes virtual environment and Oculus Touch controllers provide simple and intuitive yet powerful and detailed hands simulation which enables grabbing of 3D objects and steering of the wheelchair.

5   Conclusion

In the paper, we presented the challenges facing inclusive design and discussed the need for designers to understand the environment from the perspective of people with disabilities. We have demonstrated in that case study the influence that Immersive Virtual Reality can have on an architectural design process. The immersive quality of IVR provides a better sense of spatial awareness as well as a wider field of display than traditional VR. What's more, we show that thanks to IVR we are able to change the perspective from which we view the designed environment, enhancing the architects' empathy for the needs of different users of their designs.

Acknowledgements. The research reported in this paper was co-financed by the European Union from the European Social Fund in the framework of the project "Silesian University of Technology as a Center of Modern Education based on research and innovation" POWR.03.05.00-00-Z098/17. This work is part of the General Statutory Research Project BK-204/RAU2/2019 conducted at the Institute of Informatics, the Silesian University of Technology.

References

1. Afacan, Y.: Designing for an ageing population: Residential preferences of the turkish older people to age in place. In: Langdon, P., Clarkson, J., Robinson, P. (eds.) Designing Inclusive Futures, pp. 237–248. Springer, London (2008)
2. Altay, B., Demirkan, H.: Inclusive design: developing students' knowledge and attitude through empathic modelling. IJ Inclusive Educ. **18**(2), 196–217 (2014)

3. Azhar, S.: Role of visualization technologies in safety planning and management at construction jobsites. Procedia Eng. **171**, 215–226 (2017)
4. Bassanino, M., Wu, K., Yao, J., Khosrowshahi, F., Fernando, T., Skjaerbaek, J.: The impact of immersive virtual reality on visualization for a design review in construction. In: 14th IC on Information Visualization, pp. 585–589 (2010)
5. Benek, I.: Shaping of the living environment for the elderly in nursing homes. In: Branowski, B., Lewandowskiego, R.J., Niziołka, K., Królikowskiego, J. (ed.) Monografia od., pp. 33–64. Wyd. Pol. Lódzkiej, Poznań (2015)
6. Bonenberg, A., Branowski, B., Kurczewski, P., Lewandowska, A., Sydor, M., Torzyński, D., Zabłocki, M.: Designing for human use: examples of kitchen interiors for persons with disability and elderly people. Hum. Factors Ergon. Manuf. Serv. Ind. **29**(2), 177–186 (2019)
7. Branowski, B., Głowala, S., Pohl, P., Gabryelski, J., Zabłocki, M.: The ergonomic research on the manipulation space of the seniors and people with disabilities. In: Branowski, B. (ed.) Projektowanie dla seniorów i osób niepełnosprawnych Badania Analizy Oceny Konstrukcje, pp. 123–155. Politechnika Poznańska, Poznań (2015)
8. Dunston, P.S., Arns, L.L., Mcglothlin, J.D., Lasker, G.C., Kushner, A.G.: An immersive virtual reality mock-up for design review of hospital patient rooms. In: Wang, X., Tsai, J.J.H. (eds.) Collaborative Design in Virtual Environments, pp. 1–13. Springer, Dordrecht (2011)
9. Ergenoglu, A.S.: Universal design teaching in architectural education. Procedia - Soc. Behav. Sci. **174**, 1397–1403 (2015)
10. Eurostat: Accidents and injuries statistics. EU, ec.europa.eu/eurostat/ (2017)
11. Heylighen, A., der Linden, V.V., Steenwinkel, I.V.: Ten questions concerning inclusive design of the built environment. Build. Environ. **114**, 507–517 (2017)
12. Iwarsson, S., Slaug, B.: An instrument for assessing and analysing accessibility problems in housing. Housing Enabler. Studentlitteratur, Lund (2001)
13. Janusz, J.: Toward the new mixed reality environment for interior design. IOP Conf. Ser.: Mat. Sci. Eng. **471**, 102065 (2019)
14. Luck, R.: Inclusive design and making in practice: bringing bodily experience into closer contact with making. Design Stud. **54**, 96–119 (2018)
15. Maghool, S.A.H., Moeini, S.H.I., Arefazar, Y.: An educational application based on virtual reality technology for learning architectural details: challenges and benefits. IJ Archi. Res. ArchNet-IJARl **12**(3), 246–272 (2018)
16. Mayagoitia, R.E., van Boxstael, E., Wojgani, H., Wright, F., Tinker, A., Hanson, J.: Is remodelled extra care housing in england an inclusive and 'care-neutral' solution? In: Langdon, P., Clarkson, J., Robinson, P. (eds.) Designing Inclusive Futures, pp. 227–236. Springer, London (2008). https://doi.org/10.1007/978-1-84800-211-1_22
17. Mulligan, K., Calder, A., Mulligan, H.: Inclusive design in architectural practice: experiential learning of disability in architectural education. Disabil. Health J. **11**(2), 237–242 (2018)
18. Paes, D., Arantes, E.M., Irizary, J.: Irizarry: Immersive environment for improving the understanding of architectural 3D models: comparing user spatial perception between immersive and traditional virtual reality systems. Autom. Constr. **84**, 292–303 (2017)
19. Palmon, O., Sahar, M., Wiess, L., Oxman, R.: Virtual environments for the evaluation of human performance: towards virtual occupancy evaluation in designed environments (VOE). In: 11th International Conference on Computer Aided Architectural Design Research, pp. 521–528. Kumamoto, Japan (2006)
20. Piekarzewska, M., Wieczorkowski, R., Zajenkowska-Kozłowska, A.: Stan Zdrowia Ludności Polski W 2014 Roku. GUS, Warszawa (2016)

21. Portman, M., Natapov, A., Fisher-Gewirtzman, D.: To go where no man has gone before: virtual reality in architecture, landscape architecture and environmental planning. Comput. Environ. Urban Syst. **54**, 376–384 (2015)
22. de Souza, S.C., de Oliveira Post, A.P.D.: Universal design: an urgent need. Procedia - Soc. Behav. Sci. **216**, 338–344 (2016)
23. Turk, Y.A.: Planning - design training and universal design. Procedia - Soc. Behav. Sci. **141**, 1019–1024 (2014)
24. United Nations, Department of Economic and Social Affairs, Population Division: World Population Ageing 2015. United Nations (2015)
25. World Health Organization: World Report on Disability. WHO, Geneva (2011)
26. World Health Organization: World Report on Ageing and Health. WHO, Geneva (2015)

Spatio-Temporal Filtering for Evoked Potentials Detection

Michał Piela[1]([✉]), Marian Kotas[1], and Sonia Helena Contreras Ortiz[2]

[1] Silesian University of Technology, Gliwice, Poland
{michal.piela,marian.kotas}@polsl.pl
[2] Universidad Tecnológica de Bolívar, Cartagena, Colombia
scontreras@utb.edu.co

Abstract. We propose the new application of the spatio-temporal filtering (STF) method, which is a detection of visual evoked potentials applied to brain-computer interfaces (BCI). STF aims in creating a new, enhanced channel basing on the current and the neighbouring samples from all the input channels . The new channel of the better quality facilitates quick detection of visual evoked potential in the EEG recording by reducing number of averaging operations. The BCI experiments include precise information on the times the specific events took place. This feature allowed us to design very accurately the learning step which is based on generalized eigendecomposition and aims in determining the spatio-temporal filter weights. STF based algorithm allows to achieve good results for enhancement and detection of visual evoked potentials applied for brain-computer interfaces. Advantageous classification accuracies obtained with the use of combined spatial and temporal approach suggest the method can contribute to improvement of the existing solutions and stimulate development of more accurate and faster EEG based interfaces between machines and humans.

Keywords: Spatio-temporal filtering · Brain-computer interfaces · Visual evoked potentials

1 Introduction

Analysis of the electrical activity of the brain with regard to detection of evoked potentials can provide not only useful diagnostic information on the examined subject. It is also considered as very effective measure for brain computer interfaces (BCI). The foundation of this measure is based on the specific property of the brain's response to external stimuli: evoked potential is produced by the brain after occurrence of visual stimulus, provided the subject's attention was held on that stimulus (e.g. sudden flash of the observed item). In other words, brain selectively responds only to the stimuli which are consciously perceived by the observer. This property of event related potentials is unique for BCI's designers, who are provided with very reliable, physiological feature of the brain. P300

© Springer Nature Switzerland AG 2020
A. Gruca et al. (Eds.): ICMMI 2019, AISC 1061, pp. 34–43, 2020.
https://doi.org/10.1007/978-3-030-31964-9_4

was employed for the first time as a control signal in BCI by Farwell and Donchin in 1988 [3].

The most popular BCI system based on evoked potentials is so called speller, which aims in inputting letters to the computer by means of analysis of the subject's EEG. Typical user interface consists of a table containing letters and signs grouped in rows and columns [2]. Other, simpler applications utilize pictures which represent actions to be executed, like the ones used by [4] to control household appliances. Both interfaces function in a similar way, every row/column in the first example and every picture in the second example are flashed in random order, one time per block. The block can be then considered as time frame when each of the existing patterns is flashed exactly one time. The procedure requires saving accurate times when the flashes occur with assignments to the relevant events. Therefore, the brain's responses for these events can be distinguished later, at the stage of analysis and classification. The events related to the stimuli on which the user focuses attention will be called later in this paper as targets. Remaining events will be called as nontargets. Because of the intersubject variability of the evoked potential's features, like amplitude, shape and pre-eminently latency, the system requires proper learning phase prior to the regular use. BCI based on evoked potentials analysis can be helpful not only to the people with disabilities but can also constitute a good alternative for the healthy ones, especially in times when fast growing technological developments encourage to seek more convenient ways of communication with machines. However, there are many problems yet to be solved, so that the interfaces can be considered as efficient, fast and user friendly. Due to very low amplitude of the desired component, the potential is visually indistinguishable from spontaneous activity of the brain. Therefore, efficient detection requires multiple averaging of responses to the same stimulus what inevitably prolongs time of examination. This in turn has an adverse impact on users' concentration. Particularly, it pertains to the users whose ability to concentrate is diminished because of their health condition. Possible improvements can be applied twofold: by improving the techniques and algorithms and also by optimizing the user interfaces. In this article we focus on the former. Numerous algorithms based on both linear and nonlinear methods were developed over recent years [1,8,9,11,12,14,15].

In this paper we propose the application of the spatio-temporal filtering (STF) for detection of visual evoked potentials. The STF method was used previously in [5,6] to fetal electrocardiogram extraction. In Sect. 2, we describe STF method, and learning process. In Sect. 3, numerical experiments are presented and discussed. Finally, conclusions are formulated.

2 Methods

2.1 Spatio-Temporal Filtering

Spatio-temporal filtering method aims in determining the new channel of improved quality, basing on both spatial and temporal information. To compute the values of the formed channel, the method uses all input channels, taking into account not only current samples at given times, but also neighbouring samples which exact locations are determined by parameters J and tau. Tau denotes a constant shift in samples and J is tau's multiplier which determines how many of shifted samples are considered. Therefore, the number of $2J+1$ samples of every respective channel $i = 1, 2 \ldots, K$ are added to the resulting signal according to the following equation

$$f_{STF}^{(n)} = \sum_{i=1}^{K} \sum_{j=-J}^{J} c_{i,j} f_i^{(n+j\tau)} \tag{1}$$

To put it in a more succinct notation we can define following vectors:

$$\mathbf{c} = \begin{bmatrix} c_1 \\ c_2 \\ \ldots \\ c_{(2J+1)K} \end{bmatrix} \tag{2}$$

$$\mathbf{f}^{(n)} = \begin{bmatrix} f_1^{(n-J\tau)} \\ \ldots \\ f_K^{(n-J\tau)} \\ \ldots \\ f_1^{(n+J\tau)} \\ \ldots \\ f_K^{(n+J\tau)} \end{bmatrix} \tag{3}$$

where: $c_i, i = 1, 2 \ldots, (2J + 1)K$ are the coefficients calculated during learning step of the method, these values can be considered as the weights of the spatio-temporal filter. Single coefficient c_i represents the weight of the extended vector's corresponding dimension associated with given channel and shift in time. The matrix equation is as below:

$$f_{STF}^{(n)} = \mathbf{c}^T \mathbf{f}^{(n)} \tag{4}$$

2.2 Learning Phase

Learning period should estimate the set of weights c, which maximize signal to noise ratio of the enhanced channel. In order to perform efficient learning, we have to select the signal parts which contain desired component and parts representing the rest of the signal, which are treated as unwanted, noisy components. Following objective function is maximized in the process:

$$Q(c) = \frac{\frac{1}{N_A} \sum_{i=1}^{T_s} \sum_{n=s_i-(\frac{w}{2})}^{s_i+(\frac{w}{2})} \left(\mathbf{c}^T \mathbf{f}^{(n)}\right)^2}{\frac{1}{N_B} \sum_{i=1}^{T_s-1} \sum_{n=s_i+(\frac{w}{2})+f}^{s_{i+1}-(\frac{w}{2})-f} \left(\mathbf{c}^T \mathbf{f}^{(n)}\right)^2} \tag{5}$$

w is length of magnified window (number of samples in one target event), T_s is number of target events used in learning period, f is the offset in regard to the magnified window – we select all regions between magnified windows as the parts to be suppressed in the process. N_A and N_B denote total lengths, respectively of desired component signal and noise. Then we gather the desired signals in $\mathbf{A}_{(2J+1)K x N_A}$ and noise in $\mathbf{B}_{(2J+1)K x N_B}$:

$$Q(c) = \frac{\frac{1}{N_A}\|\mathbf{c}^T \mathbf{A}\|^2}{\frac{1}{N_B}\|\mathbf{c}^T \mathbf{B}\|^2} \tag{6}$$

We apply following modifications to the matrix $\mathbf{A}$ in the numerator: firstly, we calculate the mean value of all N_A samples, therefore the dimension is changed from N_A to 1 and instead of the matrix $\mathbf{A}$ we obtain the vector $\mathbf{g}_{(2J+1)K x 1}$. Secondly, we form analogical matrix $\mathbf{A}'_{(2J+1)K x 5N_A}$, which contains nontarget components. The number of nontarget samples is 5 times larger than of targets, it can be explained by the fact that in described experiment 6 events are considered (one is a target, five remaining are nontargets). In the next step we calculate the mean value of $5N_A$ nontarget samples, as a result we get the vector $\mathbf{g}'_{(2J+1)K x 1}$. Finally, we subtract two vectors $\mathbf{g}$ and $\mathbf{g}'$ and the resulting vector $\mathbf{h}$ is set in the final equation:

$$Q(c) = \frac{N_B\|\mathbf{c}^T \mathbf{h}\|^2}{\|\mathbf{c}^T \mathbf{B}\|^2} = \frac{N_B \mathbf{c}^T \mathbf{h}\mathbf{h}^T \mathbf{c}}{\mathbf{c}^T \mathbf{B}\mathbf{B}^T \mathbf{c}} = \frac{\mathbf{c}^T \mathbf{R}_h \mathbf{c}}{\mathbf{c}^T \mathbf{R}_B \mathbf{c}} \tag{7}$$

The function $Q(c)$ is maximized by searching for its stationary points:

$$\frac{\partial Q(\mathbf{v})}{\partial \mathbf{v}} = 0 \tag{8}$$

$$\frac{\mathbf{v}^T \mathbf{R}_B \mathbf{v}(2\mathbf{R}_h \mathbf{v}) - \mathbf{v}^T \mathbf{R}_h \mathbf{v}(2\mathbf{R}_B \mathbf{v})}{(\mathbf{v}^T \mathbf{R}_B \mathbf{v})^2} = 0 \tag{9}$$

$$\mathbf{R}_h \mathbf{v} = \frac{\mathbf{v}^T \mathbf{R}_h \mathbf{v}}{\mathbf{v}^T \mathbf{R}_B \mathbf{v}} \mathbf{R}_B \mathbf{v} = Q(\mathbf{v})\mathbf{R}_B \mathbf{v} \tag{10}$$

The final equation can be regarded as the generalized eigenproblem formula:

$$\mathbf{R}_h \mathbf{v} = \lambda \mathbf{R}_B \mathbf{v} \tag{11}$$

As a result we obtain generalized eigenvalues λ_i and corresponding eigenvectors $\mathbf{v}_i$. To maximize signal to noise ratio, we search for the eigenvector with the highest eigenvalue and this vector constitutes the final filter coefficients.

Consecutive steps of the learning phase are presented in Fig. 1.

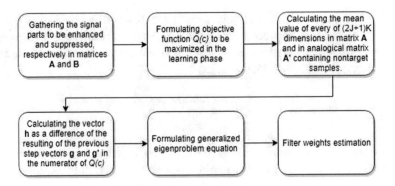

Fig. 1. Block diagram of the algorithm.

3 Numerical Experiments

The experiments described in this section aimed in investigation of the STF method performance and comparison with the methods proposed in [4]. Section 3.1 presents quantitative results. Qualitative results in the shape of visual representation of targets detection can be found in Sect. 3.2. For the purpose of testing we use the database created by [4]. The database originally contained 32-channel data of 9 subjects (4 healthy and 5 disabled with varying abilities to communicate and control of muscles). In the final set the data of subject 5th was excluded by its authors, due to lack of evidence the user understood properly the instructions given prior to the experiment start. For all other subjects complete data is available. It is split on 4 sessions, every session involves six runs (each run pertains to the individual picture the user was focused on). One run in turn consists of 20 events (20 targets and 100 nontargets). The original database was sampled with 2048 Hz, in order to lower the computational cost, we decimate it by the factor of 4. Our pre-processing step involves also initial filtering: the third order bandpass Butterworth filter is applied (by means of Matlab's function: filtfilt), cut-off frequencies are set to 1 Hz and 12 Hz. In order to compare performance of our algorithm to the results obtained using Bayesian Linear Discriminant Analysis (BLDA) method, we select the same 8 channels and apply the methodology of testing as in [4]. Considering the database lacks original separation for the learning and testing data subsets, we perform a k cross validation: all the combinations of 3 sessions are chosen as the learning datasets, the ones left out are used for testing. We apply averaging over consecutive blocks of events. To this end, we assign the values obtained after filtering for every event in the block. Having six values, which corresponds to six events (5 nontargets and 1 target) we can select one with the highest value, which constitutes the result after one block. The averaging is performed iteratively for the current block and the preceding ones. Therefore, the first average is calculated after the second block. The following one is the average of third, second and first block. The process continues this way until the last available block in the run is used.

3.1   Quantitative Results

In Figs. 2 and 3 are presented results for the disabled and healthy subjects respectively. Black lines apply to the results obtained using STF method and grey ones to the results for BLDA method.

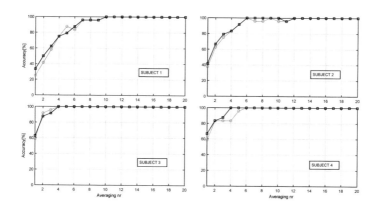

Fig. 2. Results for disabled subjects. Grey plots: BLDA, black plots: STF.

By applying STF, we obtained competitive results to BLDA, except for the subject 7. Its EEG signal contains large segments of noise what has an adverse impact on the quality of learning phase. Subject 6 does not achieve the highest accuracy what can be explained by the fact he was focusing on the wrong picture during one of runs. We obtain better results for subjects 4, 8 and 6.

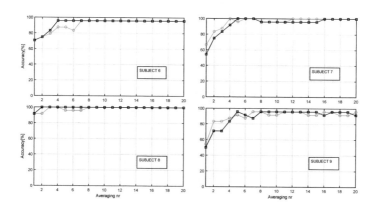

Fig. 3. Results for healthy subjects. Grey plots: BLDA, black plots: STF.

Table 1 shows accuracies averaged over all examined subjects after consecutive blocks (1–20). We can notice that in the first half of the table STF method is on average slightly advantageous than BLDA. It shows that it performs better for lower number of blocks available to average. The accuracy values never reach 100 percent, due to the fact in case of two subjects (6 and 9) none of the method has managed to reach the highest accuracy for every session these subjects took part in.

Table 1. The accuracies obtained after consecutive averaging steps, averaged over all subjects, determined using STF and BLDA methods.

Method	Accuracies averaged over all subjects [%]																			
	1	2	3	4	5	6	7	8	9	10	11	12	13	14	15	16	17	18	19	20
STF	58.9	76.0	82.3	91.1	95.3	96.9	97.9	98.4	97.9	98.4	97.9	98.4	98.4	98.4	98.4	98.4	99.0	99.0	99.0	98.4
BLDA	57.8	76.6	82.8	89.6	93.2	93.8	97.4	97.4	97.4	97.4	97.9	99.0	99.0	98.4	98.4	99.0	99.0	98.4	98.4	99.0

In order to give a better idea of amplitude of the desired component, Fig. 4 shows the overlay of short 2s EEG record (channel 1) and the averaged target response over this channel. On the right side the averaged plot is zoomed in. These exemplary plots pertain to subject 8, whose averaged target responses are the highest of all examined patients. It might be caused by the fact, this subject was highly motivated during the recording sessions (according to the annotations given by the authors of the database). Nevertheless, amplitude of the averaged plot is still rather small in comparison to normal activity of the brain. It illustrates how challenging task is to distinguish target and nontarget events taking into account only small number of averaged responses.

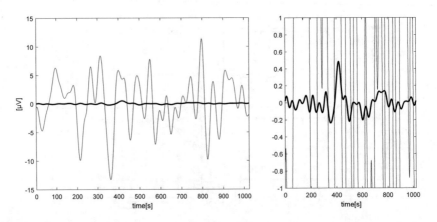

Fig. 4. On the left: 2s EEG record of subject 8 (channel 1) and the target response averaged over 4 sessions for channel 1 (bold) ; on the right: zoomed target response from the left picture.

3.2 Qualitative Results

Aim of this section is to present visually the outcome of STF method. STF filter will be called as generalized matched spatio-temporal filter (GMSTF), applied to process one channel only is simplified to generalized matched filter (GMF). Figure 5 uppermost picture shows 15s one channel EEG test signal (of subject 8). Black arrows mark positions were target events are expected. Below is the output of generalized matched filter, applied to this channel only. Finally, the lowermost picture shows results of generalized matched spatio-temporal filter, applied to 16 selected channels (out of 32 channels available in the database). Both filter outcomes were raised to the power of 3. GMF and GMSTF were designed basing on the learning part of the dataset (1 channel and 16 channels respectively). Simple algorithm was used to detect the highest peaks within every block (consecutive blocks are separated with black vertical lines), these peaks are marked with black circles. GMSTF performs well on the presented segment, only one false detection occurs. On the contrary, the performance of GMF filter is poor, it does not allow to detect any of the targets within the analyzed region. Superior efficiency of GMSTF filtering confirms advantage of spatio-temporal approach. Further development of the algorithm can include combination with the methods of time-averaging [7,10,13].

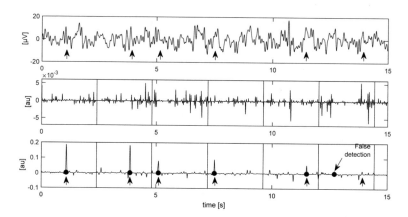

Fig. 5. Uppermost: Exemplary test signal (subject 8, channel 1), Middle: Results of generalized matched filter (GMF), Lowermost: Results of generalized matched spatio-temporal filter (GMSTF). Black arrows: expected positions of target events, black circles: detected positions of target events (the highest peaks within blocks).

4 Conclusion

Filtering of EEG signal and classification of targets/nontargets events is usually performed using linear methods of which particularly common is linear discriminant analysis and its extensions (FLDA, BLDA). Despite aforementioned methods enable to achieve good results, there is still significant room for improvement to make the BCI systems quicker, more efficient and more comfortable for end users. Emphasis is put on reducing number of averaging steps needed to detect properly the evoked potential associated to the target event. Improvements can result from adjustments of user interfaces or applying new algorithms. In this article we present the alternative approach which constitutes the new application of the spatio-temporal filtering. It allowed us to achieve competitive results to those presented in [4], in a few cases outperforming it (subjects 4, 6 and 8). Results suggest the algorithms based on STF can contribute to substantial advancements in communication between humans and machines. However, it is necessary to evaluate the method using a higher sample size to validate them.

Acknowledgement. This work was partially supported by the Ministry of Science and Higher Education funding for statutory activities of young researchers (BKM-RAu-3/2018).

References

1. Akram, F., Han, M., Kim, T.: An efficient word typing P300-BCI system using amodified T9 interface and random forest classifier. Comput. Biol. Med. **56**, 30–36 (2015)
2. Blankertz, B., Müller, K., Curio, G., Vaughan, T., Schalk, G., Wolpaw, J., Schlögl, A., Neuper, C., Pfurtscheller, G., Hinterberger, T., Schröder, M., Birbaumer, N.: The BCI competition 2003: progress and perspectives in detection and discrimination of EEG single trials. IEEE Trans. Biomed. Eng. **51**(6), 1044–1051 (2004)
3. Farwell, L.A., Donchin, E.: Talking off the top of your head: toward a mental prosthesis utilizing event-related brain potentials. Electroencephalogr. Clin. Neurophysiol. **70**, 510–523 (1988)
4. Hoffmann, U., Vesin, J., Ebrahimi, T., Diserens, K.: An efficient P300-based brain-computer interface for disabled subjects. J. Neurosci. Meth. **167**, 115–125 (2008)
5. Kotas, M., Blaszczyk, J., Moron, T.: Spatio-temporal FIR filter for fetal ECG extraction. Int. J. Inf. Electron. Eng. IACSIT Press **5**(1), 10–14 (2015)
6. Kotas, M., Jezewski, J., Horoba, K., Matonia, A.: Application of spatio-temporal filtering to fetal electrocardiogram enhancement. Comput. Meth. Prog. Biomed. **104**, 1–9 (2011)
7. Kotas, M., Pander, T., Leski, J.: Averaging of nonlinearly aligned signal cycles for noise suppression. Biomed. Signal Process. Control **21**, 157–168 (2015)
8. Krusienski, D., Sellers, E., Cabestaing, F., Bayoudh, S., McFarland, D., Vaughan, T., Wolpaw, J.: A comparison of classification techniques for the P300 speller. J. Neural Eng. **3**(4), 299–305 (2006)
9. Krusienski, D., Sellers, E., McFarland, D., Vaughan, T., Wolpaw, J.: Toward enhanced P300 speller performance. J. Neurosci. Meth. **167**(1), 15–21 (2008)
10. Leski, J.: Robust weighted averaging of biomedical signals. IEEE Trans. Biomed. Eng. **49**(8), 796–804 (2002)
11. Lotte, F., Bougrain, A., Cichocki, A., Clerc, M., Congedo, M., Rakotomamonjy, A., Yger, F.: A review of classification algorithms for EEG-based brain-computer interfaces: a 10-year update. J. Neural Eng. **15**(3), 031005 (2018)
12. Lotte, F., Congedo, M., Lecuyer, A., Arnaldi, Q.B.: A review of classification algorithms for EEG-based brain-computer interfaces. J. Neural Eng. **4**, R1–R13 (2007)

13. Momot, A., Momot, M., Leski, J.: Bayesian and empirical bayesian approach to weighted averaging of ECG signal. Tech. Sci. **55**(4) (2007)
14. Waytowich, N., Lawhern, V., Bohannon, A., Ball, K., Lance, B.: Spectraltransfer learning using information geometry for a user-independent brain-computer interface. Front. Neurosci. **10**, 430 (2016)
15. Zhang, Y., Zhou, G., Jin, J., Zhao, Q., Wang, X., Cichocki, A.: Sparse Bayesian classification of EEG for brain-computer interface. Neural Netw. Learn. Syst. **27**(11), 2256–2267 (2016)

A Review on the Vehicle to Vehicle and Vehicle to Infrastructure Communication

Krzysztof Tokarz[(✉)]

Institute of Informatics, Silesian University of Technology Gliwice, Gliwice, Poland
ktokarz@polsl.pl

Abstract. This paper presents the literature review on wireless technologies that could be used for Vehicle to Vehicle and Vehicle to Infrastructure communication. It presents the current and emerging technologies together with the results of the research made to prove the usability of network technologies in demanding vehicular data communication.

Keywords: Vehicle-to-vehicle · Vehicle-to-infrastructure · Vehicular network

1 Introduction

In the latest years, advances in the development of electronic equipment and in information technologies have created the ability to build novel Advanced Driver Assistance Systems (ADAS) applications able to identify hazardous situations and warn the driver accordingly [4]. To achieve this both internal equipment of the car and various ideas of sending and receiving drive parameters data between vehicles or between vehicles and the infrastructure have been created. The main goal of the vehicle to vehicle (V2V) communication is to implement the application of collision avoidance [18], roadway reservation [28], and autonomous intersection management [25] by allowing vehicles in transit to send data about their position and speed to others. These data can be used to at least lower risk of an accident by warning the driver who can take proper actions to avoid the danger or automatically perform preemptive actions such as braking to slow down directly by the vehicle. The automotive developers and producers are undergoing a technological revolution, manufacturing vehicles equipped with electronic devices for sensing, networking, and data processing. This leads to improve user experience, helps the drivers to drive safe and finally creates the possibility to develop autonomous vehicles [33]. In this article, the different methods, protocols, and standards used in the wireless data transfer used in modern vehicles are presented.

2 Wireless Technologies

2.1 Non-standard Data Transmission

In the literature, the idea of establishing communication between vehicles is not new. In 2002 the concept of the Ad-hoc Wireless Network (AWN) able

© Springer Nature Switzerland AG 2020
A. Gruca et al. (Eds.): ICMMI 2019, AISC 1061, pp. 44–52, 2020.
https://doi.org/10.1007/978-3-030-31964-9_5

to transmit data using a 220 MHz band on the distance up to 5 km has been presented. The authors proposed to send the information about normal driving parameters like acceleration, brake usage, turning and also emergency alerts like slippery road, ABS deployment, a tire burst, airbag activation [34]. Presented AWN solution does not need any network infrastructure located along the roads. A different approach has been proposed in 2003 in [39]. The authors describe the testbed of the system using a 2,4 GHz ISM band with Remote Central Station responsible for the coordination of data packets. Unfortunately, the range of the transmission is limited to 30 m in open space. There are also other applications using dedicated 2,4 GHz radio transmitters [19] to provide the communication channel. In [3] the author presented the system that measures a distance to the forthcoming car and generates the warning message. It is useful in the situation when the forthcoming car is not equipped with a compatible system and displays the warning message on LCD located at the rear window of the car. The author proposes additionally to send alert messages using WiFi modules.

2.2   Data Transmission with WiFi Standard

Usage of IEEE802.11a/b/g/n (WiFi) standard transmission in vehicle to vehicle (V2V) and vehicle to infrastructure (V2I) communication has been widely tested. Although results presented in [30] show that such communication is possible, achieving about 30 m of the distance, the important problem in utilizing WiFi networks as the wireless technology for vehicles is the procedure of establishing the connection. Wireless devices should be associated with each other before exchanging messages. If both nodes are stationary or when the vehicles travel in the same direction, so their relative speed is small, the time to associate is sufficient. In the dynamic situation, when the vehicles travel at high speed in opposite directions the time when nodes remain within communication range is short. In WiFi medium access model, to establish connections with the access point clients have to go through association and authentication procedures, which in the best conditions takes 600 ms [42], but normally takes several seconds [27]. Longer delays can be observed especially in dense regions or heavy traffic conditions [16]. Such amount of time might not be enough to get associated and to successfully exchange the messages [42]. When vehicles are in movement, limited range of communication influence the necessity of fast re-connecting. The Carrier Sense Multiple Access/Collision Avoidance (CSMA/CA) method, as used in 802.11n WiFi protocol, does not guarantee channel access before a fixed deadline and so it doesn't ensure time critical message transmission. The problem of fast connection between vehicles in the movement has been investigated in [42] where comparison of three popular wireless technologies (ZigBee, WiFi, and Bluetooth) has been done. Results obtained show that the devices using WiFi technology are capable to exchange data at the relative speed not exceeding 140 km/h. For V2I network, when vehicle pass the series of stationary access points (AP) the mobile station leaves and enters the footprints of the APs very frequently [20,22]. The main challenge is the handover procedure that allows switching the connection of mobile node between following APs. Paper [6] describes the analysis of handover procedure behavior in the presence of redundant APs that allows using multiple base stations simultaneously. Authors of [32] present the improvement of standard handover procedure predicting which AP should be used as the next to connect to. The proposed solution uses GPS coordinates to localize the vehicle, nearest AP and the direction of movement. Three mechanisms of predictive handover have been presented also in [12]. Authors compare standard passive method where the client starts to find new AP after losing the connection, active scanning performed before a connection is lost, and scripted handover based on information about APs locations, showing that the scripted method allows faster network reconnection.

2.3 Data Transmission with DSRC Standard

To make the idea of connected cars operable all actors must use the same technology. An open-source protocol for wireless communication named Dedicated Short Range Communications (DSRC) has been defined [40]. In October 1999, the US Federal Communication Commission allocated the 75 MHz of bandwidth at the 5.9 GHz band for DSRC-based Intelligent Transportation System (ITS) applications and adopted basic technical rules for DSRC operations [11]. The whole band is divided into seven channels where one channel is used for the transfer of safety messages, another one channel for urgent messages, and five channels are intended for non-safety applications [15]. Modeling and measurements of the 5,9 GHz radio channel for V2V communication has been described in [36] and [2].

As the name says the DSRC protocol is intended for the communication on short distances (up to 300 m) that gives the ability to transmit data between vehicles or between vehicles and infrastructure at urban areas or some chosen points on the roads. Equipment in the DSRC comprises two kinds of units: On-Board Unit (OBU) and Roadside Unit (RSU) An OBU is a transceiver that is normally mounted in or on a vehicle while RSU is a transceiver that is mounted along a road or pedestrian passageway [11]. Standard supports a speed of vehicles exceeding 200 km/h, it provides a theoretical bandwidth from 6 up to 27 Mbps [8].

The benefits of DSRC technology are low cost, ease of deployment, and supporting V2V communication as the ad-hoc networks [33].

2.4 Data Transmission with C-ITS Standard

At the same time when the DSRC standard has been developed in the US, in Europe, the work has been performed that resulted in the definition of C-ITS (Cooperative Intelligent Transportation Systems) standard. Both protocol stacks of DSRC and C-ITS are similar. Standards are based on the same frequency band 5.9 GHz the individual radio channels use a bandwidth of 10 MHz [17]. In Europe, 30 MHz (3 channels) is the primary frequency band dedicated for safety and traffic efficiency applications, 20 MHz (2 channels) band is intended for non-safety applications [14]. The band and channels in C-ITS and DSRC are similar, so it is possible to build compatible systems, but it needs the common standardization process to define the function of every channel. The Media Access Control (MAC) layer uses the same medium access CSMA/CA method with prioritization EDCA (Enhanced Distributed Channel Access) to prevent the occurrence of collisions [41]. C-ITS defines the ad-hoc routing method based on geographical coordinates for addressing and forwarding messages.

2.5 Data Transmission with Cellular Networks

The evolution of cellular networks caused the increase of network capacity and transmission speed. Cellular networks are constantly emerging that leads to possibility of using this technology as the transmission medium for vehicles. Currently available 4G LTE (Long Term Evolution) and LTE-A (Long Term Evolution - Advanced) network offer high data rate, controlled QoS, low latency and what is very important wide coverage and compatibility [33]. LTE networks can achieve the throughput up to 150 Mbit/s using dual stream transmission mode. Research presented in [31] shows that it is suitable for objects moving with the speed achieving 200 km/h. LTE-A supports downlink data rates of higher speeds from 300 Mbps to 3 Gbps [38]. What is essential for the V2V communication, especially in dangerous situations, is the transmission latency which, although small, for vehicle communication is expected 5 times reduced compared with the current 4G network [26]. Although 4G technology supports device-to-device (D2D) communication model, the current LTE D2D solution is not able to meet the low latency [29] and high reliability requirements, as it has been originally developed for public safety services with not demanding latency and reliability requirements [21].

3GPP (3rd Generation Partnership Project) proposed the evolution from LTE standard to 5G NR (New Radio). The support to vehicle communication begins with the LTE-V2X that is assumed to be the first phase of C-V2X (Cellular Vehicle to Everything) [9,35]. C-V2X allows communication over the mobile network using standard LTE technology and, even if the cellular coverage is not present direct communication between vehicles with the LTE-Direct technology.

The promising technology that has been deployed on a large scale worldwide which could fulfill vehicle communication requirements is 5G. With respect to 4G, the 5G technology is expected to achieve much better performance, including 1000 times higher system capacity, 10–100 times higher number of connecting devices [13]. The concept of device-to-device communication has been incorporated in the 5G cellular network [24].

3 Comparison of Technologies

WiFi (802.11a/b/g/n) standard has been defined for users that are stationary or moving with the low speed like while walking. This caused the selection of the MAC layer protocols that introduce significant delay while connecting or re-connecting to the network, as it was shown previously and are not suitable for fast moving devices. This standard is the base for defining IEEE802.11p (DSRC) designed for use in V2V and V2I communication. DSRC is designed to exchange data without the need of waiting for the association and authentication procedures, so devices may start to send or receive data as soon as they switch to the communication channel [27].

The comparison of 802.11a with 802.11p made in [15] shows that the latter achieves better performance measured as packet delivery ratio, throughput, and end-to-end delay. Authors of [37] presented a comparison of using WiFi and DSRC technologies for non-safety transmission using TCP protocol with single-path and multi-path mode. Results show that handover time in WiFi networks is longer than in DSRC.

Although the DRSC technology achieves better link parameters than WiFi, it is found to be still not sufficient suffering from high collision probabilities especially in increased vehicle density scenarios. This affects even more important broadcast frames, used for the emergency signaling. Lack of handshaking and acknowledgment mechanisms results in an unreliable broadcast emergency service [1]. Additionally, the DSRC technology suffers from scalability issues, unbounded delays, and lack of deterministic QoS support. The allocated DSRC frequency spectrum has limited bandwidth not sufficient to meet the high traffic demand for ultra-reliable delay-sensitive V2V safety applications [23].

Although the cellular 4G network is technically totally different concept concerning the DSRC technology, the usability of both for vehicular communication can be compared. A comparative analysis of the two technologies based on chosen common performance characteristics has been done in [33]. The comprehensive comparison of WiFi, DSRC, UMTS, LTE and LTE-A technologies for usage in vehicular communication has been presented in [5]. Both articles end with the conclusion that LTE outperforms DSRC technology but needs further improvements towards the device-to-device communication.

The DSRC technology worldwide implementation is difficult because of the incompatibility of the spectrum band, which differs depending on the region. Fortunately, utilization of the same frequency range in Europe (C-ITS) and the USA (DSRC) divided into the same 10 MHz channels makes possible to create a common standard for the hardware in vehicles [17]. Another drawback of DSRC implementation is the short range of transmission, so new, dense infrastructure should be built along the roadways. Opposite in areas not covered with cellular network signal constant operating of vehicular wireless services requires direct V2V communication that is ensured in DSRC protocol.

Basing on the comparison of issues of DSRC and cellular technologies the important requirements for 5G technology have been defined in [7] including low latency, high reliability, high throughput. Authors in [10] write that physical layer of C-V2X as compared with DSRC employ more advanced techniques, like MIMO and cooperative schemes, which are expected to improve the reliability and efficiency of data transmission. This leads to obtaining higher data rate and lower latency, but authors of [29] write that even if cellular communication provides data rates higher than DSRC the low latency requirements of safety-critical vehicular applications are not met. This situation is challenging and creates the need for further development of V2V and V2I technologies.

4 Conclusion

In this paper, a literature review has been done to identify and compare possible technologies and standards that can be used in vehicle-to-vehicle and vehicle-to-infrastructure data communication. The review shows that current works are based on the existing IEEE802.11 standard or on the emerging 5G cellular networks. Because 802.11a/b/g/n used in WiFi networks has been investigated as not sufficient so the extensions have been proposed in 802.11p in US and C-ITS in Europe. Available cellular networks based on LTE technology also has been extended with the possibility of device-to-device communication that plays an important role in emergency situations. Currently, there is no single technology available that covers the needs of reliable and fast data exchange but 5G technology is the most promising one.

Acknowledgements. This work was part of the scientific internship in the research project no. POIR.01 .01.01-00-1398/15 entitled "Development of innovative technologies in the field of active safety, which will be used in advanced driver assistance systems (ADAS) and autonomous driving systems intended for mass production" and supported by the National Centre for Research and Development in the years 2016–2020 in Poland.

The work presented in this paper was partially supported by Statutory Research funds of Institute of Informatics, Silesian University of Technology, Gliwice, Poland (grant No BK/204/RAU2/2019).

References

1. Abboud, K., Omar, H.A., Zhuang, W.: Interworking of DSRC and cellular network technologies for V2X communications: a survey. IEEE Trans. Veh. Technol. **65**(12), 9457–9470 (2016). https://doi.org/10.1109/TVT.2016.2591558

2. Acosta-marum, G., Ingram, M.A.: Doubly selective vehicle-to-vehicle channel measurements and modeling at 5.9 GHz. In: Proceedings of the International Symposium on Wireless Personal Multimedia Communications, San Diego, CA (2006). https://doi.org/10.1.1.324.3149

3. Alanezi, M.A.: A proposed system for vehicle-to-vehicle communication: low cost and network free approach. Indian J. Sci. Technol. **11**(12) (2018). http://www.indjst.org/index.php/indjst/article/view/121337

4. Anaya, J.J., Ponz, A., García, F., Talavera, E.: Motorcycle detection for ADAS through camera and V2V communication, a comparative analysis of two modern technologies. Expert. Syst. Appl. **77**, 148–159 (2017). https://doi.org/10.1016/j.eswa.2017.01.032. http://www.sciencedirect.com/science/article/pii/S0957417417300416

5. Araniti, G., Campolo, C., Condoluci, M., Iera, A., Molinaro, A.: LTE for vehicular networking: a survey. IEEE Commun. Mag. **51**(5), 148–157 (2013). https://doi.org/10.1109/MCOM.2013.6515060

6. Balasubramanian, A., Mahajan, R., Venkataramani, A., Levine, B.N., Zahorjan, J.: Interactive WiFi connectivity for moving vehicles. In: Proceedings of the ACM SIGCOMM 2008 Conference on Data Communication, SIGCOMM 2008, pp. 427–438. ACM, New York (2008). https://doi.org/10.1145/1402958.1403006. http://doi.acm.org/10.1145/1402958.1403006

7. Boban, M., Manolakis, K., Ibrahim, M., Bazzi, S., Xu, W.: Design aspects for 5G V2X physical layer. In: 2016 IEEE Conference on Standards for Communications and Networking (CSCN), pp. 1–7 (2016). https://doi.org/10.1109/CSCN.2016.7785161

8. Boussoufa-Lahlah, S., Semchedine, F., Bouallouche-Medjkoune, L.: Geographic routing protocols for vehicular ad hoc networks (VANETs): a survey. Veh. Commun. **11**, 20–31 (2018). https://doi.org/10.1016/j.vehcom.2018.01.006. http://www.sciencedirect.com/science/article/pii/S2214209616300183

9. Chen, S.z., Kang, S.l.: A tutorial on 5G and the progress in China. Front. Inf. Technol. Electron. Eng. **19**(3), 309–321 (2018). https://doi.org/10.1631/FITEE.1800070

10. Cheng, X., Zhang, R., Yang, L.: Wireless-vehicle combination: advanced PHY techniques in VCN, pp. 41–85. Springer, Cham (2019). https://doi.org/10.1007/978-3-030-02176-4_3

11. Dedicated Short Range Communications (DSRC) Service: Dedicated Short Range Communications (DSRC) Service (2019)

12. Deshpande, P., Kashyap, A., Sung, C., Das, S.R.: Predictive methods for improved vehicular WiFi access. In: Proceedings of the 7th International Conference on Mobile Systems, Applications, and Services, MobiSys 2009, pp. 263–276. ACM, New York (2009). https://doi.org/10.1145/1555816.1555843. http://doi.acm.org/10.1145/1555816.1555843

13. Dong, P., Zheng, T., Yu, S., Zhang, H., Yan, X.: Enhancing vehicular communication using 5G-enabled smart collaborative networking. IEEE Wirel. Commun. **24**(6), 72–79 (2017). https://doi.org/10.1109/MWC.2017.1600375

14. Festag, A.: Standards for vehicular communication-from IEEE 802.11p to 5G. e & i Elektrotechnik und Informationstechnik **132**(7), 409–416 (2015). https://doi.org/10.1007/s00502-015-0343-0

15. Fitah, A., Badri, A., Moughit, M., Sahel, A.: Performance of DSRC and WiFi for intelligent transport systems in VANET. Procedia Comput. Sci. **127**, 360–368 (2018). https://doi.org/10.1016/j.procs.2018.01.133. http://www.sciencedirect.com/science/article/pii/S1877050918301455, Proceedings of the First International Conference on Intelligent Computing in Data Sciences, ICDS2017

16. Fonseca, A., Vazão, T.: Applicability of position-based routing for VANET in highways and urban environment. J. Netw. Comput. Appl. **36**(3), 961–973 (2013). https://doi.org/10.1016/j.jnca.2012.03.009. http://www.sciencedirect.com/science/article/pii/S1084804512000768

17. Fuchs, H., Hofmann, F., Löhr, H., Schaaf, G.: Vehicle-2-X. In: Winner, H., Hakuli, S., Lotz, F., Singer, C. (eds.) Handbook of Driver Assistance Systems: Basic Information, Components and Systems for Active Safety and Comfort, pp. 1–17. Springer, Cham (2014). https://doi.org/10.1007/978-3-319-09840-1_28-1

18. Gehrig, S.K., Stein, F.J.: Collision avoidance for vehicle-following systems. IEEE Trans. Intell. Transp. Syst. **8**(2), 233–244 (2007). https://doi.org/10.1109/TITS.2006.888594

19. Ghatwai, N.G., Harpale, V.K., Kale, M.: Vehicle to vehicle communication for crash avoidance system. In: 2016 International Conference on Computing Communication Control and Automation (ICCUBEA), pp. 1–3 (2016). https://doi.org/10.1109/ICCUBEA.2016.7860118

20. Hasan, S.F., Siddique, N., Chakraborty, S.: Wireless technology for vehicles, pp. 1–17. Springer, Cham (2018). https://doi.org/10.1007/978-3-319-64057-0_1

21. Hu, L., Eichinger, J., Dillinger, M., Botsov, M., Gozalvez, D.: Unified device-to-device communications for low-latency and high reliable vehicle-to-x services. In: 2016 IEEE 83rd Vehicular Technology Conference (VTC Spring), pp. 1–7 (2016). https://doi.org/10.1109/VTCSpring.2016.7504518

22. Jansons, J., Petersons, E., Bogdanovs, N.: Vehicle-to-infrastructure communication based on 802.11n wireless local area network technology. In: 2012 2nd Baltic Congress on Future Internet Communications, pp. 26–31 (2012). https://doi.org/10.1109/BCFIC.2012.6217975

23. Karagiannis, G., Altintas, O., Ekici, E., Heijenk, G., Jarupan, B., Lin, K., Weil, T.: Vehicular networking: a survey and tutorial on requirements, architectures, challenges, standards and solutions. IEEE Commun. Surv. Tutor. **13**(4), 584–616 (2011). https://doi.org/10.1109/SURV.2011.061411.00019

24. Kombate, D., Wanglina: The internet of vehicles based on 5G communications. In: 2016 IEEE International Conference on Internet of Things (iThings) and IEEE Green Computing and Communications (GreenCom) and IEEE Cyber, Physical and Social Computing (CPSCom) and IEEE Smart Data (SmartData), pp. 445–448 (2016). https://doi.org/10.1109/iThings-GreenCom-CPSCom-SmartData.2016.105

25. Lee, J., Park, B.: Development and evaluation of a cooperative vehicle intersection control algorithm under the connected vehicles environment. IEEE Trans. Intell. Transp. Syst. **13**(1), 81–90 (2012). https://doi.org/10.1109/TITS.2011.2178836

26. Lianghai, J., Liu, M., Weinand, A., Schotten, H.D.: Direct vehicle-to-vehicle communication with infrastructure assistance in 5G network. In: 2017 16th Annual Mediterranean Ad Hoc Networking Workshop (Med-Hoc-Net), pp. 1–5 (2017). https://doi.org/10.1109/MedHocNet.2017.8001639

27. Liu, K., Ng, J.K.Y., Lee, V.C.S., Son, S.H., Stojmenovic, I.: Cooperative data scheduling in hybrid vehicular ad hoc networks: VANET as a software defined network. IEEE/ACM Trans. Netw. **24**(3), 1759–1773 (2016). https://doi.org/10.1109/TNET.2015.2432804

28. Liu, K., Son, S.H., Lee, V.C.S., Kapitanova, K.: A token-based admission control and request scheduling in lane reservation systems. In: 2011 14th International IEEE Conference on Intelligent Transportation Systems (ITSC), pp. 1489–1494 (2011). https://doi.org/10.1109/ITSC.2011.6082959

29. Mahmood, A., Butler, B., Sheng, Q., Zhang, W.E., Jennings, B.: Need of ambient intelligence for next-generation connected and autonomous vehicles: principles, technologies and applications. In: Guide to Ambient Intelligence in the IoT Environment. Computer Communications and Networks, pp. 133–151. Springer (2019). https://doi.org/10.1007/978-3-030-04173-1_6

30. Matsumoto, A., Yoshimura, K., Aust, S., Ito, T., Kondo, Y.: Performance evaluation of IEEE 802.11n devices for vehicular networks. In: 2009 IEEE 34th Conference on Local Computer Networks, pp. 669–670 (2009). https://doi.org/10.1109/LCN.2009.5355054

31. Merz, R., Wenger, D., Scanferla, D., Mauron, S.: Performance of LTE in a high-velocity environment: a measurement study. In: Proceedings of the 4th Workshop on All Things Cellular: Operations, Applications, & Challenges, All Things Cellular 2014, pp. 47–52. ACM, New York (2014). https://doi.org/10.1145/2627585.2627589. http://doi.acm.org/10.1145/2627585.2627589

32. Mouton, M., Castignani, G., Frank, R., Engel, T.: Enabling vehicular mobility in city-wide IEEE 802.11 networks through predictive handovers. Veh. Commun. **2**(2), 59–69 (2015). https://doi.org/10.1016/j.vehcom.2015.02.001. http://www.sciencedirect.com/science/article/pii/S2214209615000108

33. Muhammad, M., Safdar, G.A.: Survey on existing authentication issues for cellular-assisted V2X communication. Veh. Commun. **12**, 50–65 (2018). https://doi.org/10.1016/j.vehcom.2018.01.008. http://www.sciencedirect.com/science/article/pii/S2214209617302267

34. Ozguner, U., Ozguner, F., Fitz, M., Takeshita, O., Redmill, K., Zhu, W., Dogan, A.: Inter-vehicle communication: recent developments at Ohio state university. In: Intelligent Vehicle Symposium, 2002, vol. 2, pp. 570–575 (2002). https://doi.org/10.1109/IVS.2002.1188013

35. Salvatori, E.: 5G and car-to-x key technologies for autonomous road transport. ATZelektronik Worldw. **11**(6), 26–31 (2016). https://doi.org/10.1007/s38314-016-0083-x

36. Sen, I., Matolak, D.W.: Vehicle-vehicle channel models for the 5-GHz band. IEEE Trans. Intell. Transp. Syst. **9**(2), 235–245 (2008). https://doi.org/10.1109/TITS.2008.922881

37. Singh, P.K., Sharma, S., Nandi, S.K., Nandi, S.: Multipath TCP for V2i communication in SDN controlled small cell deployment of smart city. Vehicular Communications **15**, 1–15 (2019). https://doi.org/10.1016/j.vehcom.2018.11.002. http://www.sciencedirect.com/science/article/pii/S2214209618301049

38. Stoumpis, G., Karabetsos, S., Nassiopoulos, A.: An experimental framework for studying LTE and LTE-advanced. In: Proceedings of the 19th Panhellenic Conference on Informatics, PCI 2015, pp. 275–280 (2015). https://doi.org/10.1145/2801948.2801985

39. Kim, T.M., Choi, J.W.: Implementation of inter-vehicle communication system for vehicle platoon experiments via testbed. In: SICE 2003 Annual Conference (IEEE Cat. No.03TH8734), vol. 3, pp. 3414–3419 (2003)

40. West Conshohocken, PA: Standard specification for telecommunications and information exchange between roadside and vehicle systems—5-GHz band dedicated short-range communications (DSRC), medium access control (MAC), and physical layer (PHY) specifications. Technical report, West Conshohocken, PA (2018). https://doi.org/10.1520/E2213-03R18

41. Wu, X., Li, J., Scopigno, R.M., Cozzetti, H.A.: Insights into possible VANET 2.0 directions. In: Campolo, C., Molinaro, A., Scopigno, R. (eds.) Vehicular ad hoc Networks: Standards, Solutions, and Research, pp. 411–455. Springer, Cham (2015). https://doi.org/10.1007/978-3-319-15497-8_15

42. Yan, G., Rawat, D.B.: Vehicle-to-vehicle connectivity analysis for vehicular ad-hoc networks. Ad Hoc Netw. **58**, 25–35 (2017). https://doi.org/10.1016/j.adhoc.2016.11.017. http://www.sciencedirect.com/science/article/pii/S1570870516303274, Hybrid Wireless Ad Hoc Networks

Artificial Intelligence and Knowledge Discovery

Classifying Relation via Piecewise Convolutional Neural Networks with Transfer Learning

Yuting Han[1], Zheng Zhou[1], Haonan Li[2], Guoyin Wang[1(✉)], Wei Deng[1], and Zhixing Li[1]

[1] Chongqing Key Laboratory of Computational Intelligence,
Chongqing University of Posts and Telecommunications, Chongqing 400065, China
wanggy@cqupt.edu.cn
[2] The University of Melbourne, Melbourne, VIC, Australia

Abstract. Relation classification is an important semantic processing task in natural language processing (NLP). Traditional works on relation classification are primarily based on supervised methods and distant supervision which rely on the large number of labels. However, these existing methods inevitably suffer from wrong labeling problem and may not perform well in resource-poor domains. We thus utilize transfer learning methods on relation classification to enable relation classification system to adapt resource-poor domains along with different relation type. In this paper, we exploit a convolutional neural network to extract lexical and syntactic features and apply transfer learning approaches for transferring the parameters of convolutional layer pre-training on general-domain corpus. The experimental results on real-world datasets demonstrate that our approach is effective and outperforms several competitive baseline methods.

Keywords: Relation classification ·
Piecewise convolution neural networks · Transfer learning

1 Introduction

Relation classification is a crucial NLP task and plays a key role in various domains, e.g., information extraction, question answering etc. The task of relation classification is to classify semantic relations between pairs of marked entities in given texts and can be defined as follows: given a sentence S with the annotated pairs of entities e_1 and e_2, we aim to identify the relations between e_1 and e_2. For instance, in the sentence *"Mental [illness]e_1 is one of the biggest causes of personal [unhappiness]e_2 in our society"*, the entities *illness* and *unhappiness* are of relation *Cause-Effect(e_1, e_2)*.

Most existing supervised relation classification approaches rely on large amount of labeled relation-specific training data, requiring employing human

© Springer Nature Switzerland AG 2020
A. Gruca et al. (Eds.): ICMMI 2019, AISC 1061, pp. 55–65, 2020.
https://doi.org/10.1007/978-3-030-31964-9_6

to annotate the free text with information, which is time consuming and labor intensive. The erroneous annotated relation type may hurt the accuracy of results. Furthermore, the trained classifier needs readjustment when adapting the new relationship mention. Different from supervised approaches, methods based on distant supervision [1] have been developed to alleviate the costly human annotation. They assume that any sentence that contains a pair of entities expresses a relation in knowledge bases (KBs), then all sentences which contain these entity-pairs will express corresponding relation. For example, $(DonaldTrump, own, TrumpPlaza)$ is the relation fact in KB. Distant supervision will regard all sentences containing this entity-pair as the relation of own. This is much easier to automatically label training data and mainly relies on multi-instance to solve the noisy data problem. However, when people intend to obtain some new relation classification in source-poor domain, distantly supervised methods are insufficient to handle the situation.

Transfer learning is an appropriate idea to utilize the knowledge learned from resource-rich domains to transfer to resource-poor target domains with new relation type. Traditional transfer learning methods [2] can be classified into four types: instance-transfer, feature representation-transfer, parameter-transfer relational and knowledge-transfer. Parameter-transfer is adopted in this work to initialize the model parameters in the target domain. We assume that different relation types in the source and target domains share same features spaces, some words and common syntactic structures. Table 1 shows these examples. Actually, the target domain learned some distributions from source domain by transferring the parameters. The invariance of CNN structures are these distributions determined by parameter values.

Table 1. Examples of similar syntactic structures across different domain and relation types. The first and second entities are shown in italic and bold, respectively.

Syntactic pattern	Domains	Relation instance	Relation type
arg-1 of arg-2	Source domain	A *herd* is a large group of **animals**	**Memer-Collection**
		The *chapter* of the **book**	**Component-Whole**
	Target domain	*Leader* of a minority **government**	**Employ-Executive**
		The youngest *son* of **Suharto**	**Person-Family**

In this paper, we propose a syntax transferring piecewise convolutional neural networks (ST-PCNN) for resource-poor domains relation classification. The architecture is illustrated in Fig. 1. The main contributions of our work can be summarized as follows: (1) To address the relation classification in resource-poor

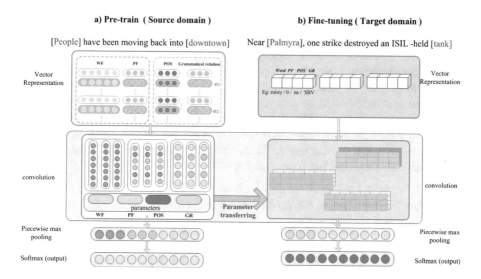

Fig. 1. The architecture of ST-PCNN

domains, we apply transfer learning approaches for transferring the parameters of model learned in one domain to new domains. (2) We explore the convolutional neural network to extract lexical and syntactic features which make full use of context information compared with the existing neural relation classification model. The experimental results show that our model achieves significant improvements in relation classification.

2   Related Work

Relation classification is a crucial task in the NLP community. Different existing approaches have been widely utilized for relation classification, such as unsupervised relation discovery and supervised classification. In the unsupervised paradigms, contextual information between entities is used as semantic relation of all entity pairs to make clustering effectively [3]. In the supervised paradigm, relation classification is considered as a multi-classification problem and feature-based, kernel-based methods are proposed to deal with the issue [4,5]. However, a comprehensible drawback of supervised method is the performance particularly depends on the quality of acquire features which are generated by NLP tools or human designed. Recent years, many researchers use deep neural networks to learn the features automatically [6,7]. These methods still suffer from a lack of sufficient labeled-data for training. Thus the distant supervision (DS) method was presented. Contrarily, the DS method is confronted with the problem of wrong labeling and noisy data, which substantially hurt the results of classification. Attention mechanism and adversarial training are proposed to alleviate the noisy problem. Furthermore, some researchers applied the transfer learning approaches for the resource-poor domains relation classification [8].

3 Methodology

Given a static source task T_S related to relation extraction, target task T_T for relation classification with $T_S \neq T_T$, and a set of sentences $\{s_1, s_2, \ldots, s_n\}$ on T_T with annotated head and tail entities e_1, e_2, we would classify each relation r between two entities. Moreover, we improve the performance by transferring the parameters of model. Figure 1 shows the ST-PCNN architecture and the processing steps are as follows. First, pre-train the ST-PCNN model and retain the convolution layer parameters. Then fine-tune the model using target data and classify the relation between entities.

3.1 The Neural Network Architecture

The neural network architecture component for relation classification is described in Fig. 1. It illustrates the procedure in four main parts, including *Vector Representation, Convolution, Piecewise Max Pooling* and *Softmax Output*. We make full use of four types of information to map each word into a low dimension feature vector for relation classification.

(1) **Word Embeddings.** Each word in a given sentence is mapped to a k dimensional real-valued vector by looking up the embedding matrix. Given a sentence consisting of N words $S = \{w_1, w_2, \ldots, w_n\}$, every word w_i is represented by a real-valued vector v_i. The dimension d^a of word embedding is x, we represent the word features (WFs) as WFs $= [v^a]$.

(2) **Position Embeddings.** Similar to [6], we use position features (PFs) to specify entity pairs. In this paper, the PFs are defined to the combination of relative distances from the current word to head and tail entities. We assume that the size of position embedding dimension d^b is y and obtain the distance vectors d_1^b and d_2^b by looking up the position embedding matrixes, transforming the relative distances of current word to e_1 and e_2 into real valued vectors, and PFs $= [v_1^b, v_2^b]$.

(3) **Part-of-Speech tags (POS).** To make use of the specific information from words themselves, we label each input word with its POS tag, e.g., noun, verb, etc. In our experiment, we only make use of a coarse-grained POS category, containing 36 different tags. We represent the POS feature vector as d^c. The dimension dc of POS tags is x.

(4) **Grammatical relations.** Dependency paths (DP) are most informative to determine the two entities relation. We utilize dependency paths to obtain the grammatical relations between words. As Fig. 2 shows, in the dependency path, each two neighbor words like *Financial* and *stress* are linked by a dependency relation *amod*. In our experiment, grammatical relations are grouped into 53 classes, mainly based on a coarse-grained classification.

Finally, we concatenate the word embeddings, position embeddings, POS and Grammatical relations embeddings of all words and transform an instance into a matrix $S \in \mathbb{R}^{s*d}$, where s is the sentence length and $d = d^a + d^b + d^c + d^e$. The matrix S is subsequently fed into the convolution layer.

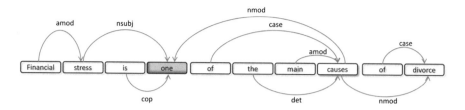

Fig. 2. Dependency parse example between stress and divorce entity pairs.

Convolution is defined as an operation between a vector of convolution weights w and a vector of inputs treating as a sequence q. The weights matrix w is considered as the filter of convolution. In the example shown in Fig. 1, we assume the length of the sliding window, filter size is $w(w = 3)$; thus , $w \in \mathbb{R}^n (n = w*d)$. Let us consider S as a sequence $\{q_1, q_2, ., q_n\}$, and let $q_{i:j}$ refer to the concatenation of q_i to q_j. We use n filters $(W = \{w_1, w_2, ., w_n\})$. Thus the convolution operation obtain another sequence $p \in \mathbb{R}^{(n+w-1)}$ expressed as follows:

$$p_{ij} = w_i q_{j-w+1:j} \ (1 \leq i \leq n \qquad 1 \leq j \leq n + w - 1) \tag{1}$$

out-of-range input values q_i, where $i < 1$ or $i > n$, are taken to be zero. The convolution result is a matrix $C = \{c_1, c_2, ., c_s\} \in \mathbb{R}^{s*(n+w-1)}$. Figure 1 shows an example of 4 different filters in the convolution procedure.

In traditional Convolution Neural Networks (CNNs), single max pooling is not sufficient to capture the fine-grained feature and structural information between two entities. In relation extraction, an input sentence can be divided into three segments based on the two selected entities. Therefore we use the piecewise max pooling [9]. As shown in Fig. 1, the output of each convolutional filter c_i is divided into three segments $\{c_{i1}, c_{i2}, c_{i3}\}$ by head and tail entities. The piecewise max pooling procedure can be expressed as follows:

$$p_{ij} = max(c_{ij}) \ (1 \leq i \leq n \qquad 1 \leq j \leq 3) \tag{2}$$

We obtain a 3-dimensional vector $p_i = \{p_{i1}, p_{i2}, p_{i3}\}$, which concatenates as $p_{1:n}$. Finally the piecewise max pooling outputs a vector:

$$g = \tanh(p_1 : n) \tag{3}$$

To compute the confidence of each relation, the feature vector g is fed into a soft max classifier.

$$o = w_1 g + b \tag{4}$$

$W_1 \in \mathbb{R}^{m*d}$ is the transformation matrix, and $o \in \mathbb{R}^m$ is the final output of the network, where m is equal to the number of possible relation types for the relation extraction system.

The ST-PCNNs based relation classification method proposed here could be stated as a quintuple $\theta = (D, W, W_1)$. (D present the concatenation of the WFs, PFs, POS and Grammatical relations embeddings.) Given an input example s, the network with parameter θ outputs the vector o, where the i-th component o_i contains the score for relation i. To obtain the conditional probability $p(i|x, \theta)$, we apply a softmax operation over all relation types:

$$p(i|x, \theta) = \frac{e^{o_i}}{\sum_{k=1}^{m} e^{o_k}} \qquad (5)$$

Given all our (suppose T) training examples $(x^{(i)}; y^{(i)})$, we can then write down the log likelihood of the parameters as follows:

$$J(\theta) = \sum_{i=1}^{T} \log p(y^{(i)}|x^{(i)}, \theta) \qquad (6)$$

We maximize $J(\theta)$ through stochastic gradient descent over shuffled mini-batches with the Adadelta [9] update rule.

3.2 Transferring Knowledge of Parameters

Parameter-transfer approaches assume that individual models for related tasks share some parameters or prior distributions of hyperparameters. Our parameter transferring procedure consists of the following steps, which we shows in Fig. 1: (a) General-domain ST-PNCC pre-training; and (b) target task ST-PCNN fine-tuning. Pre-training can accelerate the convergence rate of the model and can be beneficial for tasks with small datasets even with hundreds of labeled examples. We pre-train the ST-PCNN model on these datasets and retain the convolution layer parameters separately for use on the target relation classification task. The parameters transferred from the source domain will make the target domain data operate faster as it only needs to adapt to the idiosyncrasies of the target data. We utilize the *discriminative fine-tuning* and *slanted triangular learning rates*[10] for fine-tuning the model.

4 Experiments

Our experiments are designed to demonstrate that parameter transferring methods can increase the performance of specific domain relation classification tasks and take full advantage of automatically learned features using ST-PCNNs. We evaluate the performance of our method on several widely used datasets. We use resource-rich datasets, SemEval, NYT and FewRel Train as Source domains and three resource-poor datasets as target domains. The relations in the source and target data are basically different. Table 2 describes the division of groups The numbers of sentences and relation types describe the source domains data are rich and target domains are pool. We use AUC (Area Under Curve) as the evaluation metric in our experiments and the official macro-averaged F1-score to evaluate the model with other state-of-art models performance.

Table 2. Summary of datasets

Datasets domains	Group A		Group B		Group C	
Source target	SemEval	Dstl	NY times	SemEval	FewRel train	FewRel test
#Relation types	19	11	18,252	9	90	10
#Sentences	10,710	1200	522,611	10,710	63,000	7,000

4.1 Experimental Settings

In this paper, we use the word2vec tool to train the word embeddings on the source and target domain datasets. Our experiments directly utilize 50-dimensional vectors by comparing the word embeddings beyond the scope of this paper. We experimentally study the effects of the parameters on the model. Following previous research work, we tune our models using three-fold validation on the training set. In Table 3 we show all parameters used in the experiments.

Table 3. Summary of datasets

Window size l	3
Word dimension d^a	50
Position dimension d^b	$2*5$
POS dimension d^c	10
Grammatical dimension d^e	10
Batch size B	160
Learning rate λ	0.5
Dropout probability p	0
Sentence_max_len	120

4.2 Comparison with Other Relation Classification Methods

To evaluate the proposed method, we select the following four relation classification methods for comparison through F1-Score evaluation: **SVM** [11] is a hand-crafted features method and uses SVM for classification. **PCNN** [9] proposes a novel model dubbed the Piecewise Convolutional Neural Networks (PCNNs) with multi-instance learning to address the relation classification problem. **SDP-LSTM** [12] presents a neural network to classify the relation of two entities in a sentence. The architecture leverages the short dependency path (SDP). **HATT** [13] proposes hybrid attention-based prototypical networks for the problem of noisy few-shot relation classification. We implement them with the source codes released by the authors. Table 4 shows the F1 score for each method. We can observe that: (1) feature-based methods use the handcraft rules so it did not fit with other datasets. Thus we use the results presented by the author. (2) On the FewRel dataset, ST-PCNN significantly outperforms all other methods.

It demonstrates that the source domain dataset, FewRel training data, is more similar to the target FewRel test data thought they have different relation type. (3) On the SemEval 2010 datasets, our method's mediocre performance may be because the source data NYT noise influence the transferring result. Thus we can see the source data noise and the similarity between source and target domains data would influence the experiment results to some extent.

Table 4. Comparison of relation classification systems.

Classifier	Dstl F1	SemEval 2010 F1	FewRel F1
SVM	–	0.822	–
PCNN	0.794	0.840	0.854
SDP-LSTM	0.792	0.836	0.844
HATT	0.804	0.810	0.849
ST-PCNN	0.803	0.838	0.872

4.3 Effect of Transfer Learning

To evaluate the effects of the transfer learning method for relation classification, we empirically show the target domains datasets AUC scores with different epochs. We compare the S-PCNN and transferring results (ST-PCNN). We see the performance of transfer learning method on three target datasets. On Dstl, we observe a similarly dramatic improvement by 0.85 compared to the Dstl dataset directly running in the model with the AUC value 0.82. On FewRel, we improved by 0.94, rather than the baseline of 0.925. And on the dataset of SemEval 2010, we did not get a desired result. We see the transferring result has a slight decline compared with the baseline may because the source domain NYT corpus itself include noise data with automatic annotation. While these noise data with wrong label would also transfer to the target domain. Thus we see the source domain datasets noise would influence the final result of target data. We summarise the results in Fig. 3.

4.4 Effect of Different Features

Considering transferring different parameters corresponding to the features extracted from neural network, we performed transferring these features on the target domain datasets from the lexical and syntactic features of Table 5 to determine which type of features contributed the most. The results are presented in Table 5, from which we can observe that grammatical relations are effective for target data relation classification. The AUC score is improved remarkably when new features are added. When all of the lexical and syntactic level features are combined, we achieve relatively better results.

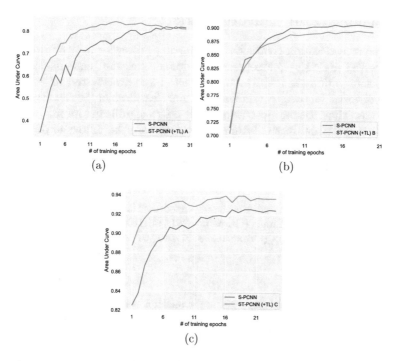

Fig. 3. AUC score for S-PCNN and ST-PCNN vs. training with different epochs on Dstl, SemEval 2010, FewRel datasets (from left to right)

Table 5. Effect of different features

Channels (Dstl)	AUC
Baseline	0.820
Word Embeddings + Position Embeddings-TL	0.834
+ POS embeddings (only)-TL	0.840
+ GR embeddings (only) -TL	0.842
+ POS + GR embeddings-TL	0.847
Channels (SemEval2010)	AUC
Baseline	0.907
Word Embeddings + Position Embeddings-TL	0.894
+ POS embeddings (only)-TL	0.894
+ GR embeddings (only) -TL	0.895
+ POS + GR embeddings-TL	0.896
Channels (FewRel)	AUC
Baseline	0.925
Word Embeddings + Position Embeddings-TL	0.938
+ POS embeddings (only)-TL	0.940
+ GR embeddings (only) -TL	0.942
+ POS + GR embeddings-TL	0.942

5   Conclusion and Future Works

In this paper, we exploit Syntactic Transferring Piecewise Convolutional Neural Networks (ST-PCNNs) for source-poor domain relation classification. In our method, source domain features are extracted and transferred to pre-train the neural networks for target domain relation classification tasks. Experimental results show that the proposed approach offers significant improvements over comparable methods. In the future, we will explore the following direction: (1) We will explore an evaluation model to measure the correlation between existing source domain data and target text corpus. (2) We will use multi-granularity convolution filter to select different granularity features.

Acknowledgements. This work was supported by the National Key Research and Development Program of China (Grant no. 2016YFB1000905), the National Natural Science Foundation of China (Grant nos. 61572091, 61772096).

References

1. Mintz, M., et al.: Distant supervision for relation extraction without labeled data. In: Proceedings of the Joint Conference of the 47th Annual Meeting of the ACL and the 4th International Joint Conference on Natural Language Processing of the AFNLP, vol. 2. Association for Computational Linguistics (2009)
2. Pan, S.J., Yang, Q.: A survey on transfer learning. IEEE Trans. Knowl. Data Eng. **22**(10), 1345–1359 (2010)
3. Min, B., et al.: Ensemble semantics for large-scale unsupervised relation extraction. In: Proceedings of the 2012 Joint Conference on Empirical Methods in Natural Language Processing and Computational Natural Language Learning. Association for Computational Linguistics (2012)
4. Culotta, A., Sorensen, J.: Dependency tree kernels for relation extraction. In: Proceedings of the 42nd Annual Meeting on Association for Computational Linguistics. Association for Computational Linguistics (2004)
5. GuoDong, Z., et al.: Exploring various knowledge in relation extraction. In: Proceedings of the 43rd Annual Meeting on Association for Computational Linguistics. Association for Computational Linguistics (2005)
6. Zeng, D., et al.: Relation classification via convolutional deep neural network (2014)
7. Nguyen, T.H., Grishman, R.: Relation extraction: perspective from convolutional neural networks. In: Proceedings of the 1st Workshop on Vector Space Modeling for Natural Language Processing (2015)
8. Liu, T., et al.: Neural relation extraction via inner-sentence noise reduction and transfer learning. arXiv preprint arXiv:1808.06738 (2018)
9. Zeng, D., et al.: Distant supervision for relation extraction via piecewise convolutional neural networks. In: Proceedings of the 2015 Conference on Empirical Methods in Natural Language Processing (2015)
10. Howard, J., Ruder, S.: Universal language model fine-tuning for text classification. arXiv preprint arXiv:1801.06146 (2018)
11. Hendrickx, I., et al.: Semeval-2010 task 8: multi-way classification of semantic relations between pairs of nominals. In: Proceedings of the Workshop on Semantic Evaluations: Recent Achievements and Future Directions. Association for Computational Linguistics (2009)

12. Xu, Y., et al.: Classifying relations via long short term memory networks along shortest dependency paths. In: Proceedings of the 2015 Conference on Empirical Methods in Natural Language Processing (2015)
13. Gao, T., et al.: Hybrid attention-based prototypical networks for noisy few-shot relation classification (2019)

Ensembles of Active Adaptive Incremental Classifiers

Michał Kozielski$^{(\boxtimes)}$ and Krzysztof Kozieł

Institute of Informatics, Silesian University of Technology,
ul. Akademicka 16, 44-100 Gliwice, Poland
michal.kozielski@polsl.pl

Abstract. An important and ubiquitous feature of the data stream generating process is its nonstationarity. Therefore, the models trained on such data streams have to be adaptive in order to react correctly on appearing concept drift. There is a number of concept drift detection methods that can be combined with a learning method to create active adaptive learner. However, none of the approaches was reported to be unambiguously the best. Within the presented work two ensemble approaches combining the adaptive base learners are applied in order to achieve higher classification quality. The analysis shows diversity of the utilised base adaptive learners justifying application of the proposed solution. The quality of results confirms that creating the ensemble of drift detectors can improve the classification quality.

Keywords: Adaptive learning · Concept drift detection ·
Data streams · Ensemble methods

1 Introduction

Amount of data generated and collected by modern computer systems poses new challenges to data analysis and processing systems. The data stream is potentially infinite, which in connection with finite resources such as time and memory has forced the adoption of new approaches to stream data processing and learning. Incremental learning [10] assumes continuous model development on incoming data examples. The quality of incremental models should aspire asymptotically to the quality of batch models, its memory requirements should not depend on a stream size and time complexity should be sub-linear.

An additional challenge for data stream analysis is imposed by the fact that, they are presumably generated in non static environments, where the concept of data generation can change [16]. This issue was analysed on general principles [8] as well as in the context of industrial applications [1,15]. There are several types of concept drift identified and addressed in literature [12] and a result of a concept drift is that an error rate of the classification or prediction model unacceptably increases. As a consequence the model becomes useless for reliable classification or prediction and the model adaptation is required.

© Springer Nature Switzerland AG 2020
A. Gruca et al. (Eds.): ICMMI 2019, AISC 1061, pp. 66–76, 2020.
https://doi.org/10.1007/978-3-030-31964-9_7

Adaptation can be performed in an active way by explicit concept drift detection that triggers a new model generation [2,3,13] or in a passive way (implicit) by continuous change of a model. Passive adaptation is achieved by creation of an ensemble, where the models are generated on consecutive data blocks [6]. The best models are retained in the ensemble and the worst (possibly the old models that do not match a new concept) are rejected.

This work focuses on the active approaches where drift detection methods are utilised. Having many methods that exhibit different characteristics and can be dedicated to different types of drift the question arises which of the methods are the best or, more generally, how to receive the best results.

Among numerous works on concept drift and adaptation three are particularly related to the given work. An analysis of the existing drift detection approaches was presented in two comparative studies [2,13]. The earlier comparison [13] was focused on nine drift detectors combined with Naive Bayes learner. The methods were compared on 12 datasets representing different types of concept drift (gradual or abrupt) and being either generated or real life ones. As a result Drift Detection Method (DDM) was pointed as the best approach. Within the later study [2] the number of analysed drift detectors was increased to 14. They were combined with one of two learners: Naive Bayes or Hoeffding Tree. The methods were compared on 7 generated data sets for which an abrupt or gradual concept drift was imposed. There were five research questions defined in the paper, however, the most general finding was that although there are differences in the quality of results when different data sets and learners are concerned the $HDDM_A$ and RDDM drift detectors were always among the best methods.

The last work [9] that should be mentioned in this place introduced stacking [17] as the solution that can combine predictions of adaptive base learners associated with different drift detectors. In this way an overall quality of results was increased. However, the approach focuses on a single drift detection method ($HDDM_A$) and it is not clearly stated what the ensemble consisted of.

The goal of this work was to verify how to receive the best classification results having a chosen classifier and a number of concept drift detectors that can produce results of various quality depending on a dataset and/or classifier. According to the best of authors' knowledge and the above-presented state of the art this task has not been thoroughly analysed so far.

The contribution of this work consists of the analysis resulting in a comparison of drift detection methods and an ensemble and meta-learning (stacking) based solutions combining the results of various methods and reducing an overall error.

The structure of the paper is as follows. Section 2 presents the issues connected with concept drift and its detection. The proposed solution is presented in Sect. 3. Section 4 presents the conducted experiments and their results. The final conclusions of the paper are enclosed in Sect. 5.

2   Concept Drift Detection

An active adaptation to concept drift, which is of interest in this work, consists of a drift detection method and a classifier. This work focuses on the drift detectors assessment. In general, drift detection methods analyse the quality of a classifier they are combined with and they trigger generation of a new classification model when the classification quality decreases.

The following drift detection methods [2,5,13] were evaluated within the presented work. The DDM (Drift Detection Method) method tracks error probability and basing on Binomial distribution properties initiates adaptation (new model generation) whenever change with a proper confidence level is registered. Achieving a lower confidence threshold triggers a *Warning* message that begins a new model generation. Achieving a higher threshold triggers a *Concept drift* message that results in a new model application. The EDDM (Early Drift Detection Method) method, unlike DDM, tracks the frequency of error, while the RDDM (Reactive Drift Detection Method) detector extends DDM discarding older instances of very long concepts. The HDDM method (Drift Detection Method based on the Hoeffding's inequality) monitors the performance of the base learner resembling DDM/EDDM but it assumes different error distribution than these methods. There are two versions of the HDDM method: $HDDM_A$ involves moving averages (more suitable for abrupt concept drift), whereas, $HDDM_W$ involves weighted moving averages (more suitable for gradual concept drift). The EWMAChartDM (Exponentially Weighted Moving Average) method applies weighted average of the older examples. The GeomAvg (Geometrical Moving Average Detection Method) method is based on the concept of assigning weights to the observations for change detection. In case of the STEPD (Statistical Test of Equal Proportions) method accuracy on recent data window is compared to overall accuracy and the difference is assessed using statistics.

The CusumDM method gives an alarm when the mean of the input data is significantly different from zero. In case of PH (Page-Hinkley Test) method, that stems from Cusum, the current and mean learner accuracy are tracked and their deviation triggers *Warning* and *Concept drift* events.

The ADWIN (Adaptive Windowing) method uses a data window of adaptive length and two sub windows to compare differences in mean values within the sub windows. The window size depends on a process stability - it is increased when no change is detected and decreased otherwise. Another methods used in this work that are based on a sliding window are the Seq2 (SeqDrift2) method and the SEED method.

3   Proposed Solution

It was shown in the previous section that there are many types of concept drift and many drift detection methods. Combining a drift detection method with an incremental learner results in an active adaptive incremental classifier, where a new model is generated and an old one is rejected when concept drift is identified.

Each of such classifiers can be applied separately to the chosen classification task. While some of the methods were reported to be more efficient in application to the chosen type of drift [2,13], it is not always clear what type of concept drift will appear in the currently analysed issue and whether it will not be a combination of several drift types. Therefore, it may not be obvious which of the adaptive approaches can be the best one.

Ensemble learning is the approach aiming to reduce prediction error by application of various base models to the given task and combination of their decisions. Within the presented work active adaptive incremental classifiers were used as the ensemble base models. Adaptation of each base model was performed independently but each base model was trained on the same data stream and each base model performed classification of the given example at the same time. The decisions of the base models were combined by means of one of two approaches in order to receive the final decision of the ensemble.

The first approach (hereinafter referred to as the *Ensemble*) derives a decision about the incoming example's class as a result of majority voting. This approach does not favour any base model, as the base model performance does not impact its vote strength.

The second approach (hereinafter referred to as the *Stacking*) uses stacking [17] to derive a decision of the ensemble. It means that an additional learner (called meta-model) is generated to combine the predictions of the base learners. The meta-model has the form of a linear model whose parameters determined during training show to what extent the final decision is based on the decisions of the chosen base model.

4   Experiments and Results

The experiments that were conducted aimed to verify several issues. The primary goal was to improve the data stream classification quality and to compare of the introduced solutions. It was also important to verify if the classification based ranking of drift detectors depends on a data set and/or base learner method. *No free lunch* assumption justifies application of ensemble learning and meta-learning, whereas stable and predictable ranking of methods would suggest application of a single best drift detector.

The experiments were carried out in the MOA tool environment [4] which offers a wide range of classification methods, drift detectors and data stream generators.

4.1   Data Sets

Within the experiments 9 data sets were used. Four of them, presented below first, were real life benchmark data sets available at MOA web page (moa.cms.waikato.ac.nz). The remaining five data sets were generated by means of stream data generators available in MOA. Each generated data set consisted of 1 000 000 examples.

Forest is a data set describing the structure of the forest represented by 30 × 30 m areas. The data were collected within the US Forest Service (USFS) Region 2 Resource Information System (RIS). Each example of the set is described by 10 numeric and 44 nominal attributes and may belong to one of 7 classes describing the type of forest area under investigation. The collection has 581 012 examples.

Airlines is a data set containing information on commercial aircraft flights in the USA between October 1987 and April 2008. The data consist of 7 attributes - 3 numeric and 4 nominal. The class of this collection is information about whether the flight was delayed or not. The set consists of 539 383 examples.

Poker is a data set containing examples of 5 cards from the deck of 52 cards for one player and evaluates the strength of his hand. Each example is described by 10 attributes (5 nominal and 5 numerical) and a decision attribute describing what is contained in a Poker Hand. The collection contains 829 201 examples.

Electricity is a data set containing examples of current electricity prices for the state of New South Wales in Australia. These prices are not fixed and depend on the demand and supply on the market. Actual data is updated every 5 min. It is predicted whether the new price will be higher or lower compared to the average price of the last 24 h. Each example in a set is described by 7 numerical attributes and 1 nominal attribute. The collection has 45 312 examples.

Agr1 and *Agr2* data sets were created by means of Agrawal generator creating the sets having 9 attributes - 6 numeric and 3 nominal, where a binary class labels are calculated by one of 10 functions. *Agr1* data set combines function 3 and function 7, whereas, *Agr2* data set combines function 10 and function 1 in order to receive data containing concept drift.

LED1 and *LED2* - are the data sets describing the display of a digit on a 7 segment LED display. The data sets consist of 24 binary attributes, where 17 are non-significant ones and the class label represents one of 10 digits that can be displayed. *LED1* data set does not contain concept drift. In case of *LED2* data set concept drift was introduced into three attributes.

RBF is a data set where examples were generated for a radial basis function using centroids. Concept drift in this data set is imposed by changing the centroid positions. The data set was generated using 40 centroids, it consists of 10 numeric attributes and 4 possible class labels.

4.2   Experimental Setup

MOA offers many evaluation and adaptation approaches, and classification methods. The evaluated solutions were compared using prequential evaluation [7, 11]. Two basic incremental classification methods were used within the experiments: Naive Bayes and Hoeffding Trees [14]. Each classification method was executed in combination with several drift detection methods creating different adaptive learning approaches. There were used 12 drift detection methods available in MOA that were described in Sect. 2: ADWIN, CusumDM, DDM, EDDM, EWMAChartDM, GeomAvg, HDDM$_A$, HDDM$_W$, PH, RDDM, SEED, STEPD, Seq2. Additionally, the NoChange approach was included into the comparison.

Finally, having the ensemble of adaptive models two approaches were implemented to derive a final decision, as described in Sect. 3. Within the *Ensemble* solution a majority voting was applied, whereas within the *Stacking* solution an incremental linear regression was applied as a meta-model to calculate base model weights and derive a combined final decision. All the methods listed above were executed with default parameter setting what resulted from the research presented in [13] where the parameter value optimisation performed for several drift detection methods resulted in the values close to the default. Furthermore, within the research presented in [2, 9] the default parameter values of drift detection methods were set.

4.3   Results

The classification accuracy of all the base learners and *Ensemble*, and *Stacking* approaches when Naive Bayes classifier was applied is presented in Table 1. The results when Hoeffding Tree classifier was applied are presented in Table 2.

Table 1. Classification accuracy of the compared adaptive learning methods (Naive Bayes used as a base learner)

	Airlines	Poker	Forest	Electricity	Agr1	Agr2	LED1	LED2	RBF
ADWIN	55.45	55.32	36.53	42.66	**71.87**	67.23	73.94	57.61	26.62
CusumDM	67.46	72.54	81.55	79.21	71.77	69.67	73.86	57.45	28.09
DDM	65.33	61.97	88.03	81.18	**71.87**	68.15	73.94	57.61	28.47
EDDM	65.18	77.48	86.08	84.83	71.86	68.31	73.94	57.61	28.57
EWMAChartDM	63.64	**79.13**	**90.16**	**86.76**	68.39	67.77	69.23	55.02	28.58
GeomAvg	64.55	59.55	60.52	73.36	**71.87**	67.95	73.94	57.61	28.58
HDDM$_A$	67.22	76.48	87.42	84.92	71.85	68.93	73.94	57.58	28.3
HDDM$_W$	65.34	77.11	86.23	84.09	70.76	69.59	72.48	53.39	26.57
PH	67.04	70.67	80.06	78.04	**71.87**	69.17	73.93	57.61	28.66
RDDM	67.49	76.67	86.86	84.19	71.78	**69.76**	73.86	57.46	27.81
SEED	66.7	75.09	84.5	82.16	71.77	69.17	61.68	55.05	27.59
STEPD	65.72	77.19	87.53	84.49	71.05	69.59	69.72	51.45	26.41
Seq2	66.59	72.25	82.44	79.68	**71.87**	67.95	73.64	57.59	28.58
NoChange	64.55	59.55	60.52	73.36	**71.87**	67.95	73.94	57.61	28.58
Ensemble	67.61	77.99	88.1	84.89	**71.87**	68.23	73.94	57.62	28.59
Stacking	**67.87**	79.04	89.42	85.72	71.84	69.31	**74.04**	**57.76**	**29.16**

Tables 1 and 2 highlight the best results with respect to classification quality. Comparing both tables it is possible to evaluate the performance of the base models. A higher classification quality is achieved when the adaptive approaches were based on Hoeffding Tree method (Table 2) in all except one issue.

Table 2. Classification accuracy of the compared adaptive learning methods (Hoeffding Tree used as a base learner)

	Airlines	Poker	Forest	Electricity	Agr1	Agr2	LED1	LED2	RBF
ADWIN	55.44	55.31	36.53	42.66	71.91	67.23	12.31	11.68	**30.91**
CusumDM	65.78	72.85	83.01	81.71	71.46	73.54	73.67	57.31	28.64
DDM	65.29	72.74	87.36	85.41	71.91	72.66	73.78	57.50	**30.91**
EDDM	65.06	77.31	86.00	84.91	71.90	73.44	73.88	57.54	**30.91**
EWMAChartDM	63.81	78.62	89.70	86.45	67.76	67.51	69.25	55.01	**30.91**
GeomAvg	65.08	76.07	80.31	79.20	71.91	73.78	73.88	57.53	**30.91**
HDDM$_A$	64.99	76.40	87.24	85.71	71.67	71.19	73.64	57.27	**30.91**
HDDM$_W$	65.02	77.12	85.97	85.06	69.83	69.12	72.46	53.34	28.11
PH	65.18	71.30	81.65	81.95	71.91	74.03	73.39	57.11	30.24
RDDM	66.00	76.70	86.42	85.18	71.11	71.10	73.66	57.28	28.65
SEED	65.43	75.19	84.47	83.54	71.36	69.76	58.12	50.77	28.94
STEPD	65.37	77.13	86.99	85.29	70.18	69.12	69.67	51.12	28.32
Seq2	65.31	72.51	82.85	82.84	71.91	73.78	71.34	56.47	**30.91**
NoChange	65.08	76.07	80.31	79.20	71.91	73.78	73.88	57.53	**30.91**
Ensemble	68.02	78.72	88.90	87.14	72.04	73.89	73.94	57.64	**30.91**
Stacking	**68.27**	**80.03**	**89.84**	**87.25**	**72.17**	**74.50**	**74.06**	**57.79**	29.53

Besides, the tables show indirectly that default configuration of some of the drift detection methods was far from perfect. The ADWIN detector reached very poor results because it either found no drift in data or adapted for each example. The GeomAvg detector, in turn, did not identify any concept drift.

In order to make a comparison of drift detection methods easier they are additionally compared in Table 3. This table contains '+' marks if a given drift detector combined with a given base learner achieved the quality of results on a given data set among the best three approaches. The top three scores are summed up for each base classification method.

The visualisation in the form of Table 3 shows that the group of the best methods can change depending on the undertaken issue and the base learner that was applied. None of the drift detectors is among the best approaches for all the analysed data sets. There are adaptive approaches which are impacted by the type of base classification model combined with a given drift detection method, e.g. DDM combined with Hoeffding Tree achieves better results than with Naive Bayes, whereas STEPD shows the opposite trend.

Finally, the rankings of drift detection methods when combined with Naive Bayes and Hoeffding Tree (Figs. 1 and 2 respectively) were created. The rankings are presented in a form of critical distance diagrams (the results from the Nemenyi post-hoc test at 0.05 significance level on the datasets on the analysed data sets).

Table 3. The number of times the performance of the drift sensor has been classified in the top three results

	Naive Bayes											Hoeffding Tree										
ADWIN											0											0
CusumDM	+				+	+					3	+					+	+				3
DDM		+									1			+	+			+	+			4
EDDM		+		+			+	+			4		+			+			+			3
EWMAChartDM		+	+	+							3		+	+	+							3
GeomAvg											0											0
HDDM$_A$	+			+	+						3			+	+	+						3
HDDM$_W$											0											0
PH						+		+			2						+			+		2
RDDM	+				+						2	+								+		1
SEED											0	+								+		2
STEPD		+	+	+	+						4	+										1
Seq2							+				1											0
NoChange				+		+	+	+			4				+	+	+	+	+			5

Figures 1 and 2 show that the introduced *Ensemble* and *Stacking* approaches were justified and achieved the best classification quality. *Stacking* is the best solution on average within the presented comparison.

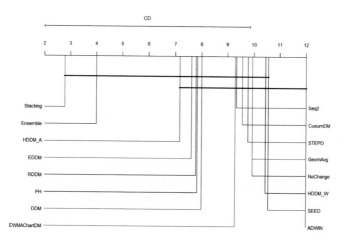

Fig. 1. Critical distance diagram ranking wrt. classification accuracy the drift detection methods combined with Naive Bayes classifier

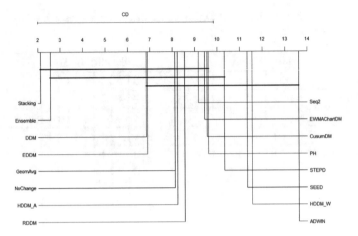

Fig. 2. Critical distance diagram ranking wrt. classification accuracy the drift detection methods combined with Hoeffding Tree classifier

5   Conclusions

Within the presented work a number of drift detection methods were evaluated in order to verify whether there is a clear leader in terms of classification accuracy. It was shown that the results and rankings vary depending on the analysed issue and base learning method. Therefore, two ensemble methods were applied, one using simple majority voting and the other one using a meta-learning approach (stacking).

The applied ensemble methods were distinguished by the highest quality among the compared adaptive approaches in accordance with the calculated rankings and *Stacking* approach achieved the best classification performance on average.

Application of ensemble approaches turned out to be beneficial also for another reason. Interestingly, some of the drift detection methods performed very poorly (their operation suggested incorrect parameter setting). This highlights another advantage of ensemble methods where poor performance of one of the ensemble members does not ruin utterly the classification abilities of the whole system.

Acknowledgements. This work was carried out within the statutory research project of the Institute of Informatics, Silesian University of Technology: BK-204/RAU2/2019.

References

1. Bakirov, R., Gabrys, B., Fay, D.: Multiple adaptive mechanisms for data-driven soft sensors. Comput. Chem. Eng. **96**, 42–54 (2017). https://doi.org/10.1016/j.compchemeng.2016.08.017. http://www.sciencedirect.com/science/article/pii/S0098135416302782

2. Barros, R.S.M., Santos, S.G.T.: A large-scale comparison of concept drift detectors. Inf. Sci. **451–452**, 348–370 (2018). https://doi.org/10.1016/j.ins.2018.04.014
3. Bifet, A., Gavaldà, R.: Adaptive learning from evolving data streams. In: Adams, N.M., Robardet, C., Siebes, A., Boulicaut, J.F. (eds.) Advances in Intelligent Data Analysis VIII, pp. 249–260. Springer, Heidelberg (2009)
4. Bifet, A., Holmes, G., Kirkby, R., Pfahringer, B.: MOA: massive online analysis. J. Mach. Learn. Res. **11**, 1601–1604 (2010). http://portal.acm.org/citation.cfm?id=1859903
5. Bifet, A., Read, J., Pfahringer, B., Holmes, G., Žliobaitė, I.: CD-MOA: change detection framework for massive online analysis. In: Tucker, A., Höppner, F., Siebes, A., Swift, S. (eds.) Advances in Intelligent Data Analysis XII, pp. 92–103. Springer, Heidelberg (2013)
6. Brzezinski, D., Stefanowski, J.: Reacting to different types of concept drift: the accuracy updated ensemble algorithm. IEEE Trans. Neural Netw. Learn. Syst. **25**(1), 81–94 (2014). https://doi.org/10.1109/TNNLS.2013.2251352
7. Dawid, A.P.: Present position and potential developments: some personal views statistical theory the prequential approach. J. R. Stat. Soc. Ser. (Gen.) **147**(2), 278–290 (1984). https://doi.org/10.2307/2981683. https://rss.onlinelibrary.wiley.com/doi/abs/10.2307/2981683
8. Ditzler, G., Roveri, M., Alippi, C., Polikar, R.: Learning in nonstationary environments: a survey. IEEE Comput. Intell. Mag. **10**(4), 12–25 (2015). https://doi.org/10.1109/MCI.2015.2471196
9. Frías-Blanco, I., Verdecia-Cabrera, A., Ortiz-Díaz, A., Carvalho, A.: Fast adaptive stacking of ensembles. In: Proceedings of the 31st Annual ACM Symposium on Applied Computing - SAC 2016, pp. 929–934 (2016). https://doi.org/10.1145/2851613.2851655. http://dl.acm.org/citation.cfm?doid=2851613.2851655
10. Gaber, M.M., Zaslavsky, A., Krishnaswamy, S.: Data stream mining. In: Maimon, O., Rokach, L. (eds.) Data Mining and Knowledge Discovery Handbook, pp. 759–787. Springer, Boston (2010). https://doi.org/10.1007/978-0-387-09823-4_39
11. Gama, J.a., Sebastião, R., Rodrigues, P.P.: Issues in evaluation of stream learning algorithms. In: Proceedings of the 15th ACM SIGKDD International Conference on Knowledge Discovery and Data Mining, KDD 2009, pp. 329–338. ACM, New York (2009). https://doi.org/10.1145/1557019.1557060. http://doi.acm.org/10.1145/1557019.1557060
12. Gama, J.a., Žliobaitė, I., Bifet, A., Pechenizkiy, M., Bouchachia, A.: A survey on concept drift adaptation. ACM Comput. Surv. **46**(4), 44:1–44:37 (2014). https://doi.org/10.1145/2523813. http://doi.acm.org/10.1145/2523813
13. Gonçalves, P.M., De Carvalho Santos, S.G.T., Barros, R.S.M., Vieira, D.C.L.: A comparative study on concept drift detectors (2014). https://doi.org/10.1016/j.eswa.2014.07.019
14. Hulten, G., Spencer, L., Domingos, P.: Mining time-changing data streams. In: Proceedings of the Seventh ACM SIGKDD International Conference on Knowledge Discovery and Data Mining, KDD 2001, pp. 97–106. ACM, New York (2001). https://doi.org/10.1145/502512.502529. http://doi.acm.org/10.1145/502512.502529
15. Ślęzak, D., Grzegorowski, M., Janusz, A., Kozielski, M., Nguyen, S.H., Sikora, M., Stawicki, S., Wróbel, L.: A framework for learning andembedding multi-sensor forecasting models into a decision support system: a case study of methane concentration in coal mines. Inf. Sci. **451–452**, 112–133 (2018). https://doi.org/10.1016/j.ins.2018.04.026. http://www.sciencedirect.com/science/article/pii/S0020025518302822

16. Widmer, G., Kubat, M.: Learning in the presence of concept drift and hidden contexts. Mach. Learn. **23**(1), 69–101 (1996). https://doi.org/10.1023/A:1018046501280
17. Wolpert, D.H.: Stacked generalization. Neural Netw. **5**(2), 241–259 (1992). https://doi.org/10.1016/S0893-6080(05)80023-1. http://www.sciencedirect.com/science/article/pii/S0893608005800231

Influence of the Applied Outlier Detection Methods on the Quality of Classification

Błażej Moska, Daniel Kostrzewa[✉], and Robert Brzeski

Institute of Informatics, Silesian University of Technology,
Akademicka 16, 44-100 Gliwice, Poland
{daniel.kostrzewa,robert.brzeski}@polsl.pl

Abstract. This paper presents a comparison of a few chosen outlier detection methods and test quality of classification, both before and after the procedure of removing outliers. Using a few selected methods of outlier detection on several selected data sets, the process of elimination of atypical data was carried out. Atypical data may be of various nature. It can be noise or can be incorrect data. However, they can also be correct data, which for some reason are different from typical data. The removal of non-typical data may have a different effect on the classification quality. It may be dependent on the used method of removing unusual data but also on the nature of used data. Therefore, the classification process was carried out on the original data as well as with the outliers removed. The obtained results were compared and discussed.

Keywords: Outlier detection method · Classification · Data analysis · Interquartile method · Distance-based method · Local outlier factor · Angle based outlier detection

1 Introduction

In classification [18, 20, 23, 35] process, one of the critical elements is the corresponding data set. That is the set large enough, containing so many samples – data vectors, that a sufficient learning process can be carried out. The set of vectors, in the classification process, will be divided into classes – sets of vectors, correspondingly similar to each other. Moreover, in the context of similarity, the question arises: What are the vectors that diverge with their similarity, to the typical representatives of a given class? On the one hand, they may be unusual but correct representatives of a given class. On the other hand, they may be noise or incorrect data, that should not be in a given set of data. So in such a situation, removing them from the data set could improve the classification process. Deleting outliers [16, 27] is a process of removing individual data vectors from the whole set, and this should not be confused with deleting individual attributes from the data vector, which takes place in data dimensionality reduction [19] process.

© Springer Nature Switzerland AG 2020
A. Gruca et al. (Eds.): ICMMI 2019, AISC 1061, pp. 77–88, 2020.
https://doi.org/10.1007/978-3-030-31964-9_8

The research was carried out on several data collections available at the UCI Machine Learning Repository [12] and BioInformatics Research Group [13]. Moreover, there may be differences in the quality of the classification, due to diversity in data sets: number of training vectors, number of outliers, number of classes, nature of data. Also, the method of detecting an outlier set of data to be removed can have an impact because various detection methods can determine a different set of outliers.

The classification process itself has been implemented using several selected classifiers [34, 35]: classification tree, kNN, logistic regression, random forest, and SVM. Probably, different classifiers deal with non-typical data individually. For some classifiers, the outlier may cause a more significant disorder, and for others, it may be neutral. Therefore, as the result of the classification, the average values of the accuracy, precision, and recall parameters obtained for these classifiers was presented.

1.1   Related Work

Outlier detection is a standard procedure in widely understood data analysis. Firstly, the definition of outlier should be given. As defined in [6], *an outlier is an observation which deviates so much from the other observations as to arouse suspicions that it was generated by a different mechanism.* The concept of observation will be used frequently and can be defined as a point located in a m-dimensional space, and each dimension is represented as a variable or equivalently column or feature. A tabular-like structure consisting of n observations with m columns represents dataset.

Detection of outliers can be made using various methods [3, 15, 27, 29], in detail, i.e., local outlier factor [8], angle-based outlier detection [21], detection based on statistical measures, clustering methods and data mining methods [2], detection based on classification [1, 32], detection based on supervised and unsupervised machine learning techniques [25]. The result of the detection process of the non-typical data [4, 33] can have many applications. Outlier detection could be applied to various disciplines. As an example, fraud detection in the financial industry could be showcased, where financial transactions are surveyed to check if any fraud occurred. Another example which could be highlighted is cybersecurity, where outlier is considered as a possible violation of security. More cases worth to be mentioned are: medical diagnosis, fault detection [24], customer segmentation, and more [27]. There are many ways to use an outlier, despite the lack of information, whether the data found are incorrect or simply unusual data. However, if there is already specified, which data is atypical, would it not be worth removing them from the set intended for data classification? Therefore, the question arises how outliers removal is influencing the classification process. To some extent, these issues have been discussed in [17, 22, 26, 30]. Moreover, there are more examples of positive affection for outlier removal. For instance, in [16] robust decision tree algorithm was proposed. In this paper, authors emphasize especially the aspect of decreasing decision tree size, thus yielding decrease in overall memory complexity and increase in generalization capability.

There is also another approach presented in [14]. The authors were focused on prototype selection for regression problems. Their method determines the best subset of input vectors to build the best possible model.

Apart from delivering useful information about underlying processes, outliers can potentially have a significant impact on machine learning models application. Whether the effect is positive or negative is contradictory. In [3], it is suggested to leave outliers prior to modelling since they give essential information, whereas in [31] authors unveil some examples where removing outliers before fitting models resulted in improved classification accuracy on testing dataset.

Generally, there is not much information on this subject. In this article, an entire set of experiments will be carried out to check how the situation looks like in practice. The quality of classification itself can, of course, be determined in many ways [5,9,28]. Choosing the right way depends on the purpose of the classification. One of the most popular and therefore easy to compare is the accuracy parameter.

1.2 Contribution

The main contribution of the paper is to compare a few chosen outlier detection methods and test quality of classification both before and after this procedure. Bearing this in mind and according to cited sources it is not directly possible to emphasize the statement that outlier removal would either improve or lower classification accuracy, this question remains open, but it will be undoubtedly valuable if the overall quality improves after the outlier removal procedure.

1.3 Paper Structure

The paper is structured as follows. Section 2 describes in details the outlier detection algorithms investigated in this paper. In Sect. 3, we present the results of our experiments, which have been carried out to evaluate the implemented methods. Finally, Sect. 4 concludes the paper and shows the main goals of our ongoing research.

2 Outlier Detection Methods

In this section, we discuss the implemented outlier detection algorithms in detail. These methods have been experimentally validated in Sect. 3.

All presented outlier detection methods work for numeric attributes. For other types of data, e.g., nominal attributes, instead of methods operating on distances, you could use, for example, domain-based methods [11] or methods that operate on a frequency of occurrences of categorical values.

2.1   Interquartile Method

Interquartile method (IQR) is the univariate method related to the concept of interquartile range, defined as follows:

$$IQR = Q_3 - Q_1, \tag{1}$$

where Q_3 is third quartile, Q_1 is the first quartile. Observation is considered as an outlier if its value is outside the following range [7, 10]:

$$[Q_1 - k \cdot IQR, Q_3 + k \cdot IQR]. \tag{2}$$

The most important advantage of the method is robustness to skewed distribution [29], but this applies only to some extent – the more skewed distribution, the more observation could be classified as outliers. Another asset of this method is the simplicity of computation, as it only requires the calculation of quartiles. This procedure is implemented in many libraries available from Python. IQR method is also convenient to illustrate graphically, as shown in Fig. 1, which could be considered as a strength in some situations.

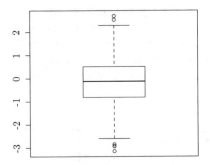

Fig. 1. An example of outliers in the normal distribution.

On the other hand, IQR performs poorly when applied to a dataset of a small number of observations. Moreover, IQR only applies to one dimension at the time, so this may be not sufficient in higher dimensions. However, it is possible to enable the user to select in how many dimensions observation should be detected as an outlier to be considered as an outlier regardless of the dimension.

2.2   Distance-Based Method

In the distance-based method (DB), observation is considered as an outlier [24] then and only then when the distance of at the most of p percent of objects in given dataset D from object O is smaller than some fixed value *dist*. It means that the outlying object has a small number of neighbours in *dist* radius. This definition can be provided in mathematical form:

$$\frac{|\{O'|d(O,O') < dist\}|}{|D|} \leq p. \tag{3}$$

The simplest solution for calculating the DB method is to perform nested loops. This would result in square computational complexity, which can potentially cause problems with a large dataset, but according to [24] in many cases real complexity is linear. The reason is that most datasets do not have many outliers, and so the inner loop does not have to be executed in full range, that is to say, for whole dataset D. This method's biggest drawback is sensitivity to appropriate choose of *dist* and p parameters.

As stated in [24], some solutions to this problem were proposed. One of them changes the definition of an outlier slightly: Suppose that $D^k(O)$ is k-nearest neighbour distance of object O in dataset D. The object O is treated as an outlier then and only then when at least p percent of the object in dataset D have lower k-nearest neighbour distance than object O. As can be observed in definition, this does not require *dist* parameter. Unfortunately, the computational complexity problem was not solved.

2.3   Local Outlier Factor

While IQR and DB methods aim at detecting global outliers, in some cases, this approach may be insufficient, since local characteristics of data are not taken into consideration. To solve this kind of task, a local approach was introduced, with local outlier factor (LOF) as a flagship method.

Firstly, let us define two concepts related to this algorithm, *k-distance* and *reachability distance*. *k-distance* is defined as follows [8,24]:

k-distance is a distance between object p and object o, such that for at least k objects o' condition $d(p,o') \leq d(p,o)$ is satisfied and for at most $k-1$ objects $o'd(p,o') < d(p,o)$ condition is satisfied. In other words, *k-distance* describes radius in which at most k objects are contained.

In [8], *reachability distance* is defined with the following formula:

$$reach - dist_k(p,o) = \max\{k - distance(o), d(p,o)\}. \tag{4}$$

The idea is illustrated in Fig. 2. What can be observed is that reachability distance is useful when a point is close enough to *k-distance* but does not lay in *k-distance* radius. This makes the choice of distance more smooth.

Other parameters, necessary to define Local Outliers Factors, are *MinPts* and *density*. *MinPts* parameter defines minimal number of observations in a given neighbourhood of point p. Combining previously mentioned parameters *local reachability density* (*lrd*) can be defined:

$$lrd_{MinPts}(p) = \frac{1}{\frac{\Sigma_{o \in N_{MinPts}(p)} reach - dist_{MinPts}(p,o)}{|N_{MinPts}(p)|}}. \tag{5}$$

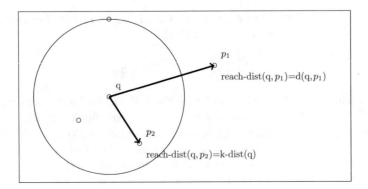

Fig. 2. The idea of reachability distance.

Finally, using described definitions, a Local Outlier Factor is defined with the following equation:

$$LOF_{MinPts}(p) = \frac{1}{\frac{\sum_{o \in N_{MinPts}(p)} \frac{lrd_{MinPts}(o)}{lrd_{MinPts}(p)}}{|N_{MinPts(p)}|}}. \tag{6}$$

The expression can be interpreted as an average quotient of local reachability density of points belonging to the neighbourhood defined with *MinPts* to local reachability density of point p.

LOF values significantly greater than one can lead to a suspicion that the object is an outlier. If the value approximates at one, it is an indicator that the object is comparable to its neighbours, values lower than one implies that the object is an inlier. In work [24], it is stated that LOF can be calculated in two steps. Firstly, *MinPts* neighbourhood of all dataset points are calculated. This requires calculation of nearest neighbour distance for all n object in dataset, which can be performed in $O(nlogn)$ time.

2.4   Angle Based Outlier Detection

Previous two methods perform calculations using some distance metric. While this approach is effective in the space of low dimensionality, it can fail when the number of dimensions is huge, because of what is known in the literature as a "curse of dimensionality" [21], which can be illustrated by simplified formula:

$$\lim_{d \to \infty} \frac{dist_{max} - dist_{min}}{dist_{min}} \to 0, \tag{7}$$

where d is the number of dimensions.

This means that with a growing number of dimensions, the difference between furthest and nearest point vanishes.

The general concept behind Angle Based Outlier Detection is illustrated in Fig. 3. It can be observed that the variability of angles is lower when the

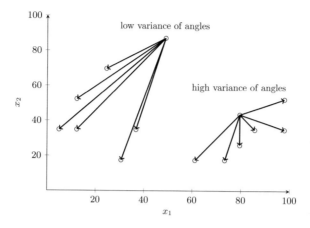

Fig. 3. General idea of angle based outlier detection.

point is an outlier. In this method, distance is calculated only as a purpose of normalization.

Variability of angles is a variance of angles from one point to all pairs of other points weighted by their distance [21]. The dot product divided by norm is the cosine of an angle:

$$\cos(\alpha) = \frac{x \cdot y}{|x| \cdot |y|}. \tag{8}$$

The lower the variance of cosines, the lower the variance of angles.

In the algorithm for each observation in a dataset, all pairs of points must be considered, resulting in cubic computational complexity $O(n^3)$ [21], which is barely computable on large datasets. This is a severe drawback. On the other hand, contrary to many other methods, angle based outlier factor does not need any parameters. As an output algorithm returns a vector of variances of cosines, which can be left as a rank, but one can specify threshold value, below which observations are considered as outliers.

3   Experiments

The quantitative experiments were conducted on datasets available online on the UCI Machine Learning Repository [12] and BioInformatics Research Group [13] websites. Detailed information about input data is provided in Table 1.

The methodology of the experiments carried out is as follows. In the beginning, the classifiers (i.e., k-nearest neighbours, logistic regression, classification tree, random forest, and support vector machine) are trained using the training sets without removing the outliers. 4-fold cross-validation is performed to obtain the optimal parameters for the classifiers. The best set of parameters for each dataset is chosen. Then, the first numerical experiment is performed on test sets.

The selected outlier detection algorithms are then applied on the training sets, the classifiers are trained, and the rest of numerical experiments are executed afterwards. The ABOD algorithm was not performed for training sets containing more than 1000 observations, because of the very long execution time (more than a few hours for one run of the algorithm).

Table 1. Datasets used for experiments.

Dataset	No. of classes	No. of features	No. of training observations	No. of testing observations
Cancer	14	16063	144	46
Digits	10	64	3823	1797
Madelon	2	500	2000	600
Occupancy	2	5	8143	2665
Smartphone	12	561	7767	3162
Urban	9	147	168	507

Table 2. The results of performed experiments.

Outlier detection method	Parameters	Accuracy	Precision	Recall	% of removed observations
Cancer dataset					
None	–	0.464	0.367	0.396	–
IQR	vars = 1	–	–	–	100.0
	vars = 2	–	–	–	100.0
	vars = 3	–	–	–	100.0
DB	perc = 0.9; k = 5	*0.444*	0.337	0.368	9.7
	perc = 0.9; k = 10	0.434	*0.361*	0.369	9.7
	perc = 0.95; k = 5	0.430	*0.358*	0.357	4.9
LOF	n = 5	0.414	0.338	0.343	10.4
	n = 10	0.426	0.333	0.354	10.4
	n = 15	0.428	0.343	0.360	10.4
ABOD	treshold = 0.1	*0.460*	*0.348*	*0.384*	0.0
	treshold = 0.2	0.420	0.343	0.352	4.2
	treshold = 0.3	0.296	0.197	0.217	47.2
Digits dataset					
None	–	0.832	0.891	0.830	–
IQR	vars = 1	0.736	0.834	0.736	54.3
	vars = 2	0.756	0.856	0.755	33.9
	vars = 3	0.778	**0.904**	0.779	18.5
DB	perc = 0.9; k = 5	0.810	**0.909**	0.809	10.0
	perc = 0.9; k = 10	0.806	**0.909**	0.807	10.0
	perc = 0.95; k = 5	*0.820*	**0.899**	*0.820*	5.0
LOF	n = 5	*0.826*	**0.908**	*0.826*	10.0
	n = 10	**0.832**	**0.907**	**0.830**	10.0
	n = 15	*0.830*	**0.907**	**0.832**	10.0

(*continued*)

Table 2. (*continued*)

Outlier detection method	Parameters	Accuracy	Precision	Recall	% of removed observations
Madelon dataset					
None	–	0.790	0.795	0.795	–
IQR	vars = 1	0.500	0.500	0.500	97.2
	vars = 2	0.510	0.510	0.510	90.2
	vars = 3	0.600	0.600	0.595	77.3
DB	perc = 0.9; k = 5	0.750	0.760	0.755	10.0
	perc = 0.9; k = 10	0.750	0.755	0.755	10.0
	perc = 0.95; k = 5	**0.790**	**0.795**	**0.795**	5.0
LOF	n = 5	**0.800**	**0.805**	**0.800**	10.0
	n = 10	**0.800**	**0.805**	**0.805**	10.0
	n = 15	**0.800**	**0.795**	**0.795**	10.0
Occupancy dataset					
None	–	0.91	0.924	0.901	–
IQR	vars = 1	**0.914**	0.922	0.898	1.6
	vars = 2	**0.948**	**0.947**	**0.949**	0.0
	vars = 3	**0.910**	0.918	0.897	0.0
DB	perc = 0.9; k = 5	**0.916**	**0.925**	**0.902**	10.0
	perc = 0.9; k = 10	**0.950**	**0.947**	**0.950**	10.0
	perc = 0.95; k = 5	**0.910**	0.919	0.899	5.0
LOF	n = 5	**0.930**	**0.934**	**0.924**	10.0
	n = 10	**0.934**	**0.939**	**0.929**	10.0
	n = 15	**0.934**	**0.937**	**0.930**	10.0
Smartphone dataset					
None	–	0.888	0.771	0.731	–
IQR	vars = 1	0.408	0.137	0.209	82.8
	vars = 2	0.492	0.170	0.253	77.0
	vars = 3	0.500	0.200	0.257	73.5
DB	perc = 0.9; k = 5	**0.892**	*0.762*	**0.735**	10.0
	perc = 0.9; k = 10	**0.892**	**0.773**	**0.739**	10.0
	perc = 0.95; k = 5	**0.888**	*0.764*	*0.728*	5.0
LOF	n = 5	*0.880*	0.734	*0.718*	10.0
	n = 10	*0.886*	*0.756*	*0.728*	10.0
	n = 15	*0.882*	0.744	*0.722*	10.0
Urban dataset					
None	–	0.524	0.506	0.508	–
IQR	vars = 1	0.472	0.410	0.432	44.0
	vars = 2	0.488	0.474	0.452	37.5
	vars = 3	0.494	0.482	0.456	33.3
DB	perc = 0.9; k = 5	*0.504*	0.484	0.486	9.5
	perc = 0.9; k = 10	*0.512*	*0.488*	*0.495*	9.5
	perc = 0.95; k = 5	0.498	0.479	0.477	4.8
LOF	n = 5	*0.510*	**0.508**	*0.494*	10.1
	n = 10	0.496	0.497	0.482	10.1
	n = 15	*0.516*	**0.517**	*0.499*	10.1
ABOD	treshold = 0.1	*0.522*	*0.502*	**0.508**	0.0
	treshold = 0.2	**0.532**	**0.512**	**0.516**	0.0
	treshold = 0.3	**0.528**	**0.510**	**0.518**	0.0

All quantitative outcomes are gathered in Table 2. The results, which gave better or the same values than the classification without the use of any outlier detection algorithm, are shown in bold in the table. Values that are no more than 0.02 worse than the result without removing the outliers are marked in italics, as results that do not differ significantly in quality.

4   Conclusions and Future Work

As shown in Table 2, for the Cancer data set and the IQR method it was not possible to carry out the classification, because this method removed all objects from the training set, treating them as outliers. In some cases, the effectiveness of the classification differs slightly from the expected values (this is especially visible in cases where no objects have been removed from the learning set). These differences are due to the randomness of some of the classifiers used as well as the randomness of cross-validation.

As can be seen, the results of numerical experiments do not allow to determine which method of detecting outliers is the best. However, some interesting conclusions can be drawn. The IQR method is the simplest one, but it also gives the worst results. ABOD is not suitable (due to its high computational complexity) for training sets containing a large number of observations, which is unacceptable nowadays. However, for two data sets (for which it was possible to run the algorithm) ABOD gave the best outcomes. LOF, in 4 out of 6 cases, turns out to be slightly better in terms of the quality of classification than DB, but these differences are not very large. Also, the number of removed outliers is similar in both methods.

The already drawn conclusions show that the use of LOF and DB methods brings the best results. Their application does not cause any significant deterioration in the quality of classification. In some cases, it even allows for its improvement. What is beneficial is that removing part of the observations from the training set may contribute to a reduction in the classification time of new observations.

Our current research focuses on the use of outlier detection methods for classifying music by genre. We are planning to use these methods not only to discard whole observations but also to extract features from tracks. We also want to test a modified version of the ABOD algorithm (Fast-ABOD), which calculates cosine variances but only for the k nearest neighbours of a given point. According to the authors of the method – this is not the best solution, because there is a need to calculate the distance. Perhaps thanks to this, it will be possible to apply the method for large sets (which in two cases, for which it was possible to run the algorithm, giving the best results).

Acknowledgements. This work was supported by Statutory Research funds of Institute of Informatics, Silesian University of Technology, Gliwice, Poland (BKM-556/RAU2/2018 – DK, BK/204/RAU2/2019 – RB).

References

1. Abe, N., Zadrozny, B., Langford, J.: Outlier detection by active learning. In: Proceedings of the 12th ACM SIGKDD International Conference on Knowledge Discovery and Data Mining, pp. 504–509. ACM (2006)
2. Acuna, E., Rodriguez, C.: A meta analysis study of outlier detection methods in classification. Technical paper, Department of Mathematics, University of Puerto Rico at Mayaguez, pp. 1–25 (2004)
3. Acuña, E., Rodriguez, C.: On detection of outliers and their effect in supervised classification. University of Puerto Rico at Mayaguez (2004)
4. An, W., Liang, M.: Fuzzy support vector machine based on within-class scatter for classification problems with outliers or noises. Neurocomputing **110**, 101–110 (2013)
5. Arie, B.D.: Comparison of classification accuracy using Cohen's Weighted Kappa. Expert Syst. Appl. **34**(2), 825–832 (2008)
6. Ben-Gal, I.: Outlier detection. In: Maimon, O., Rokach, L. (eds.) Data Mining and Knowledge Discovery Handbook, pp. 131–146. Springer, Boston (2005)
7. Boschetti, A., Massaron, L.: Python. Podstawy nauki o danych. Helion (2017). (in Polish)
8. Breunig, M.M., Kriegel, H.P., Ng, R.T., Sander, J.: LOF: identifying density-based local outliers. In: ACM SIGMOD Record, vol. 29, pp. 93–104. ACM (2000)
9. Costa, E., Lorena, A., Carvalho, A., Freitas, A.: A review of performance evaluation measures for hierarchical classifiers. In: Evaluation Methods for Machine Learning II: Papers from the AAAI 2007 Workshop, pp. 1–6 (2007)
10. Davis, M.: Statistics for life scientists. https://www.sfu.ca/~jackd/Stat203_2011/Wk02_1_Full.pdf. Accessed 19 Mar 2019
11. Domingues, R., Filippone, M., Michiardi, P., Zouaoui, J.: A comparative evaluation of outlier detection algorithms: experiments and analyses. Pattern Recogn. **74**, 406–421 (2018)
12. Dua, D., Graff, C.: UCI Machine Learning Repository (2017). http://archive.ics.uci.edu/ml. Accessed 19 Mar 2019
13. GCM – Global Cancer Map dataset. http://eps.upo.es/bigs/datasets.html. Accessed 19 Mar 2019
14. Guillén, A., Herrera, L.J., Rubio, G., Pomares, H., Lendasse, A., Rojas, I.: New method for instance or prototype selection using mutual information in time series prediction. Neurocomputing **73**(10–12), 2030–2038 (2010)
15. Hodge, V., Austin, J.: A survey of outlier detection methodologies. Artif. Intell. Rev. **22**(2), 85–126 (2004)
16. John, G.H.: Robust decision trees: removing outliers from databases. In: KDD, pp. 174–179 (1995)
17. Kalisch, M., Michalak, M., Sikora, M., Wróbel, L., Przystałka, P.: Influence of outliers introduction on predictive models quality. In: Kozielski, S., Mrozek, D., Kasprowski, P., Małysiak-Mrozek, B., Kostrzewa, D. (eds.) Beyond Databases, Architectures and Structures. Advanced Technologies for Data Mining and Knowledge Discovery, pp. 79–93. Springer, Cham (2015)
18. Kostrzewa, D., Brzeski, R.: Adjusting parameters of the classifiers in multiclass classification. In: International Conference: Beyond Databases, Architectures and Structures, pp. 89–101. Springer (2017)
19. Kostrzewa, D., Brzeski, R.: The data dimensionality reduction in the classification process through greedy backward feature elimination. In: International Conference on Man–Machine Interactions, pp. 397–407. Springer (2017)

20. Kostrzewa, D., Brzeski, R., Kubanski, M.: The classification of music by the genre using the KNN classifier. In: International Conference: Beyond Databases, Architectures and Structures, pp. 233–242. Springer (2018)

21. Kriegel, H.P., Zimek, A., et al.: Angle-based outlier detection in high-dimensional data. In: Proceedings of the 14th ACM SIGKDD international conference on Knowledge discovery and data mining, pp. 444–452. ACM (2008)

22. Li, W., Mo, W., Zhang, X., Squiers, J.J., Lu, Y., Sellke, E.W., Fan, W., DiMaio, J.M., Thatcher, J.E.: Outlier detection and removal improves accuracy of machine learning approach to multispectral burn diagnostic imaging. J. Biomed. Opt. **20**(12), 121,305 (2015)

23. Mehra, N., Gupta, S.: Survey on multiclass classification methods (2013)

24. Morzy, T.: Eksploracja danych. Metody i algorytmy, Wydawnictwo Naukowe PWN, Warszawa, pp. 326–327 (2013). (in Polish)

25. Omar, S., Ngadi, A., Jebur, H.H.: Machine learning techniques for anomaly detection: an overview. Int. J. Comput. Appl. **79**(2), 33–41 (2013)

26. Padmaja, T.M., Dhulipalla, N., Bapi, R.S., Krishna, P.R.: Unbalanced data classification using extreme outlier elimination and sampling techniques for fraud detection. In: 15th International Conference on Advanced Computing and Communications (ADCOM 2007), pp. 511–516. IEEE (2007)

27. Pei, J.: Outlier detection, data mining. http://www.cs.sfu.ca/CourseCentral/741/jpei/slides/Outlier%20Detection%201.pdf. Accessed 19 Mar 2019

28. Powers, D.M.: What the f-measure doesn't measure: features, flaws, fallacies and fixes. arXiv preprint arXiv:1503.06410 (2015)

29. Seo, S.: A review and comparison of methods for detecting outliers in univariate data sets. Ph.D. thesis, University of Pittsburgh (2006)

30. Smith, M.R., Martinez, T.: Improving classification accuracy by identifying and removing instances that should be misclassified. In: The 2011 International Joint Conference on Neural Networks, pp. 2690–2697. IEEE (2011)

31. Tallón-Ballesteros, A.J., Riquelme, J.C.: Deleting or keeping outliers for classifier training? In: 2014 Sixth World Congress on Nature and Biologically Inspired Computing (NaBIC 2014), pp. 281–286. IEEE (2014)

32. Upadhyaya, S., Singh, K.: Classification based outlier detection techniques. Int. J. Comput. Trends Technol. **3**(2), 294–298 (2012)

33. Weekley, R.A., Goodrich, R.K., Cornman, L.B.: An algorithm for classification and outlier detection of time-series data. J. Atmos. Ocean. Technol. **27**(1), 94–107 (2010)

34. Weka 3. http://www.cs.waikato.ac.nz/~ml/weka/. Accessed 19 Mar 2019

35. Wu, X., Kumar, V., Quinlan, J.R., Ghosh, J., Yang, Q., Motoda, H., McLachlan, G.J., Ng, A., Liu, B., Philip, S.Y., et al.: Top 10 algorithms in data mining. Knowl. Inf. Syst. **14**(1), 1–37 (2008)

Predictive Algorithms in Social Sciences – Problematic Internet Use Example

Eryka Probierz[1]([⊠]), Wojciech Sikora[1], Adam Gałuszka[1], and Anita Gałuszka[2]

[1] Institute of Informatics, Silesian University of Technology, Gliwice, Poland
erykaprobierz@gmail.com
[2] Katowice School of Economics, Katowice, Poland

Abstract. The article aims to show the possible applications of predictive algorithms in the field of social sciences. Due to the high diversity of data obtained in research, there is a need to look for solutions that allows better exploratory data analysis. The article uses the example of Problematic Internet Use, which is a phenomenon related to excessive use of the network and the negative effects associated with it. This phenomenon is the subject of psychological research, in the field of variables that can be considered as predictors, elements of construct image, or the possibility of predicting its development. The purpose of this article is to propose a method allowing to build a pre-evidential model of the occurrence of the problematic Internet Use using the possessed data, and to determine the correlations between variables constituting this phenomenon.

Keywords: Predictive algorithms · Problematic internet use ·
Social science · Complex data

1 Introduction

Predictive analytics finds more and more applications every day and its rapidly expanding branch of science. Constantly growing data sets render the use of such techniques almost indispensable. Methods used in social sciences based on studying the distribution of variables and on correlation and regression analyses have certain limitations. They can be used without problems with data mining methods that can offer more extensive and accurate analysis and interpretation results [8]. In addition, the high complexity of data from social sciences requires the use of techniques that allows shaping of multidimensional models that capture interdependencies and allow for the construction of predictions based on variables included in such models. Using classical techniques is a challenge, because the assumptions of many analyses (both before and after the actual post-hoc implementation) are not possible. In addition, the constructs and phenomena described in social sciences, for example in psychology are multifactorial phenomena requiring complex research, often dependent on situational aspects

© Springer Nature Switzerland AG 2020
A. Gruca et al. (Eds.): ICMMI 2019, AISC 1061, pp. 89–98, 2020.
https://doi.org/10.1007/978-3-030-31964-9_9

and on socio-demographic factors of the subjects. Controlling a large number of variables allows not only to obtain a greater degree of explanation of a given phenomenon but also to the possibility of making more accurate predictive analyses allowing, for example, the classification or detection of constructs in the subjects.

Knowledge discovery in this field can be particularly problematic due to characteristics of the data. High redundancy, large number of variables combined with limited amount of examples and incomplete information are only a few of the challenges. Considering the possibilities of prediction algorithms is crucial in the aspects of diagnostics, building prevention programs or classification. It allows using of high complexity data and modelling the problem as a whole, taking into account the correlations between variables. The use of such techniques has a chance to obtain results burdened with a smaller error, but also to allow for a more in-depth analysis of the problem, both in the context of the causes, the picture and possible consequences. Classical prediction models based on regression require fulfilment of many assumptions, that can not be met due to the nature of the data. The purpose of this study is to try to predict Problematic Internet Use without using questionnaire, that measure this concept. Previous authors research focused on explaining the phenomenon of problematic Internet use with use of social variables, this work takes up problem of predicting the problematic use of the Internet, without measuring the construct itself. The purpose of this study is to use and distinguish methods that allow for the pre-depiction of Problematic Internet Use. For this purpose, a database was collected. One part were variables constituting the reasons for the occurrence of Problematic Internet Use, the other part was a questionnaire, which measures the intensity of Problematic Internet Use. After checking with modelling of structural equations that the proposed variables explain the model well, the questionnaire on Problematic Internet Use was removed and an attempt was made to predict this phenomenon only on the basis of variables that constitute its cause.

2   Problematic Internet Use

2.1   Model and Aspects of Problematic Internet Use

The phenomenon of problematic Internet use is based on the assumptions of cognitive-behavioural psychology. In order to talk about the occurrence of this problem, it is important to have many overlapping factors. The key element is having access to the Internet. Age in which people started using the Internet was also significant variable. As it results from the research, the earlier a person starts using this technology, the more risk it entails for him. In addition adolescents and young adults, due to their age, constitute a window of vulnerability [1], when the risk of this phenomenon is the highest. It indicates both the generational aspect, that is, the youngest generations use the Internet from the early years and in larger quantities than their parents or grand-parents, and increase plasticity of the brain, which means the use of the Internet can have long-term effects. Another factor is certain mechanisms that maintain and rein-force problematic use of the Internet. The user develops certain beliefs which, intensified through his behaviour, lead to the consolidation and reproduction [5]. The way in which the Internet is used is also important. For this purpose, a distinction was made between the specific and non-specific use of the Internet. Specific use is most often related to another disorder and is its continuation, eg a person addicted to gambling can also do it on the Internet and a shopaholic, in addition to shopping in stationary stores, will also do online. Non-specific use is the way of using the Internet, which does not have one dominant activity, but it consists of many, smaller, inter-penetrating components [2].

2.2   Subscales

The tool to measure this phenomenon allows determining not only the intensity of occurrence of a given construct in the subjects, but also to distinguish subscales, allowing more precise determination of problems related to excessive Internet use.

One of the subscales is the preference of on-line contacts over those that can be conducted in the real world. It is a most preferred choice of conducting mediated interpersonal communication, that is with the use of technology. As it results from the research, such preference may came from previously educated patterns of proceedings or experience, which makes the person consider online communication to be more secure and less risky than the traditional one.

Using the Internet to regulate the mood is another subscale, it specifies the motives of the person using the network. The desire to improve the mood or reduce unpleasant emotions may indicate difficulties in regulating emotions in a different way, or about creating preferences for just such a way of dealing with emotions.

The next subscale is inadequate self-regulation, which may manifest itself in the cognitive and behavioral aspects. The cognitive aspect speaks of such preoccupation with the use of the network that a person using the Internet is not able to focus their attention on something else, whereas compulsive use of the Internet refers to a constant return to using the network, even if there are no obvious reasons, it can be presented by means of a reflexive checking of social networks on the Smartphone or browsing information messages while performing other activities.

The last subscale is the negative consequences that can be observed in the case of such Internet usage. They refer to such zones of human life as family work, friends and attitude to oneself [3].

3   Need for Using Algorithms in Social Sciences

The use of such methods to analyze data from social domains provides an opportunity for better predictions and a wider analysis of inter-relationships.

Advanced techniques of exploratory data analysis allow to change the current paradigm of looking at data, and thus provide a new way of analysis that offer a better understanding and modeling of specific phenomena. Although most commonly data mining is used for large data with thousands, millions and more examples, models provided by machine learning algorithms can have great utility with smaller data sets. Even with limited data those algorithms can offer very good predictive capabilities. Some models are very easy to interpret and thus give practical information about causes of studied problem, other methods can be used to divide the data in groups with similar features which can also be very useful. Moreover, there are practices that allow to mitigate problems with data such as low representation of some groups, incompleteness, scarcity or biases in the data. It appears that many problems in the classical analysis of data from social sciences, can be solved by using machine learning methods and informatics.

4   Data

The study was carried out on 543 respondents. The study involved people aged from 15 to 26. Respondents filled out a questionnaire in which they answered questions about their place of residence, average time spent on the Internet each day, the number of devices that the person uses and the type of activity they undertake on the Internet.

Self-assessment, perceived social support, personality traits and early non-adaptive schemes were proposed as variables associated with the phenomenon of Problematic Internet Use. Social support is here understood as a feeling to belonging to a given group. Level of social support was based on it's effect on self-esteem. The more perceived social support person is given, the strongest effect on self-esteem. Other factor is evaluating itself by the way that person is treated by others. The personality traits studied are: extraversion, emotional stability, openness to experience, agreeableness and conscientiousness.

The study used a Rosenberg self-assessment questionnaire [15], a tool for the study perceived social support ISEL [4], a short inventory of personality traits TIPI-PL [7], a short version examining early non-adaptive schemes YSQ-S3-SW [12] and a scale examining Problematic Using the Internet, GPIUS2 [13]. Our model of research is shown below. The data set had information on personality traits, self-assessment, age, social support and early non-adaptive schemes, as well as on the time the person spends on the Internet. In addition, the database contained information about the severity of Problematic Internet Use. As reasons, self-esteem, 4 social support subscales, 5 personality traits and 18 early non-adaptive patterns were defined.

To model picture of the phenomena in previous studies all of the predictors was checked [14]. In order to do that conducted structural equation modelling technique. Index of minimum discrepancy was equal to 3,07, goodness of fit index equals 0,99, comparative fit index equals 0,99 and root mean square error of approximation equals 0,06. All of those indexes mean that model fit to the obtained data.

An attempt was made to define high or low level of this construct, only by means of its predictors, omitting the data concerning the questionnaire examining this phenomenon directly. The main idea was to classify the level of Problematic Internet Use as high or low by variables obtained in the questionnaires, omitting the results from direct study. The ability to distinguish those two stages can be an important element of preventive and prophylaxis programs (Fig. 1).

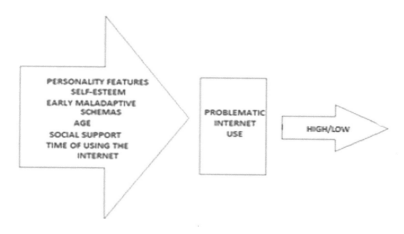

Fig. 1. Data schema

The data schema presents in the first arrow all possible predictors that have been recognized for having a directional connection with Problematic Internet Use [6,14]. Then, the subjects were also intensively examined for Problematic Internet Use with the help of a tool (Generalised Problematic Internet Use Scale 2). The purpose of the conducted analyses is to find a method that allows predicting of low or high intensity of Problematic Internet Use with the help of predictors, without the need to use a tool to examine it. We define high Problematic Internet Use as 50% above the median and low as 50% below the median (Table 1).

Table 1. Data features

Feature	
Number of people	543
Women/Men	391/152
Age: 15–17/18–20/21–23/24–26	36/240/209/58
Village/Small City/Average City/Large City	96/30/164/253
Number of used devices 1/2/>= 3	108/341/94

5   Results

In order to make analyzes, a classifier was created. One approach was to divide the classes into two based on a value that was half of the possible points. The data had an unbalanced values of classes. However, there are others techniques that help to mitigate that problem, the approach that was proposed was to divide and create parallel classes based on the median value. Balanced data are easier to model. It means that the more equal groups, the better results can be achieved using suggested methods.

Due to the exploratory nature of the research, many different methods were used. Two methods occurred best results from described below. A few different methods were used:

- K nearest neighbours
- Rule Induction
- Gradient Boosted Trees
- Naive Bayes
- Logistic Regression
- Support Vector Machine (SVM)
- Common Forest
- Decision Tree

Some of presented methods were previously used in data from social science. The most commonly used is logistic regression [10] which were used also in personal characteristic data, decision tree method were used to predict internet game addiction [9] and k nearest neighbour is commonly used in missing data imputation [11].

5.1   All Results

In the Fig. 2 is presented the summary of obtained results. Each line represents a ROC curve for each method.

As it is shown in the Fig. 2 decision trees and k-NN gave the worst results.

5.2   Best Results

The best results were obtained by logistic regression and SVM. In the Fig. 3 are presented results of ROC curve of logistic regression and two SVM models with different parameters.

Table 2 contains a comparison of the classification measures. Support Vector Machine create a linear model that is able to classify data which are not linear.

Table 3 presents the confusion matrix for logistic regression and shows the results matrix for SVM. AUC (Area Under ROC Curve) is also known as statistics and it is used for diagnosis of accuracy of quantitative tests. The satisfactory result of AUC indicates that the decision threshold and the stability of data prioritization remain unchanged. Recall also known as sensitivity is ratio that

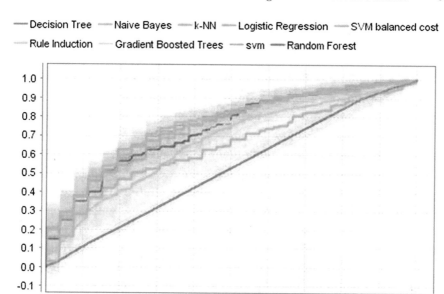

Fig. 2. Comparison of ROC (Rate of Change) curves

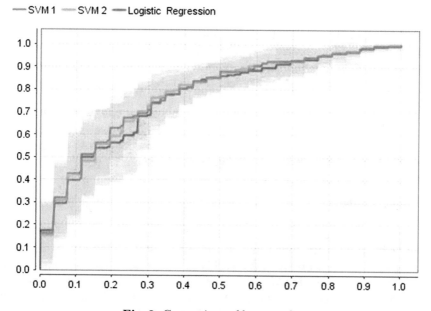

Fig. 3. Comparison of best results

Table 2. Classification measures

Indicator	Logistic Regression	SVM
Accuracy	70,54%	71,75%
Precision	72,66%	74,44%
Recall	69,34%	70,43%
AUC	0,77	0,77
f_measure	70,76%	72,08%

Table 3. Confusion matrix

Logistic regression			
	True false	True true	Class precision
pred. false	188	86	68,61%
pred. true	74	195	72,49%
class recall	71,76%	69,40%	
SVM			
pred. false	191	83	69,71%
pred. true	71	198	73,61%
class recall	72,90%	70,46%	

focuses on correctly predicted positive observations in observed class. Precision result is taking up probability of relevance retrieved document. Another indicator, accuracy, measure a ratio of correctly predicted observation from total amount of observation. The last indicator is harmonic average of the indicators: precision and recall.

The possibility of classifying people to a high or low risk group can be used as a tool in preventive and prophylaxis programs. The use of methods from machine-learning to analyze data from social domains provides an opportunity for a wider analysis of inter-relationships and predictions.

6 Discussion

Scarcity of the data, as it is one of the main problems with this kind of data, led to poor results when applying rule induction and k nearest neighbours algorithm. Decision trees prove to be unfit for problems with such limited amount of data and generated models were unuseful. Ensemble of decision trees build with gradient boosting method was slightly better but also didn't provide a satisfactory predictor. Performance measures worth attention were produced by such algorithms as Naive Bayes, Logistic regression and support-vector machine. Overall classification accuracy above 70% and AUC measure around 0,76 is quite good considering the nature of the problem and indicates that machine learning can be used as a tool in predictive and preventive programs.

7 Conclusions

The aim of this article was to show the possibilities of applying machine learning techniques to social science data. Although only a few algorithms were used, the results were diverse and some of them have proven useful. The use of such methods to analyze data from social domains provides an opportunity for better predictions and a wider analysis of inter-relationships. The obtained results may be a promising start of research that allows measuring a given construct not directly, but using variables that determine its causes. This is especially important in psychology, where there is the fear of assessment. People who receive a questionnaire, whose task is to directly check the Problematic Internet Use, can direct the answers in order to fall out better than it is in reality. Self-report questionnaires are also subject to another error. What a person thinks about himself is not always a faithful representation of reality. Increasing the variables that can predict a given phenomenon, without asking directly about it, can contribute to greater credibility of research and better detection of Problematic Internet Use. Further works are planned to extend the classifier from a two-stage class, sensitive only to borderline cases, to a four-level class comprising: low, medium low, medium high and high.

Acknowledgements. For first and second authors co-financed by the European Union through the European Social Fund (grant POWR.03.02.00-00-I029).

This work has been partially supported by Institute of Automatic Control BK Grant 02/010/BK18/0102 (BK/200/Rau1/2018) in the year 2019. The analysis has been performed with the use of IT infrastructure of GeCONiI Upper Silesian Centre for Computational Science and Engineering (NCBiR grant no POIG.02.03.01-24-099/13).

References

1. Arbaugh, W.A., Fithen, W.L., McHugh, J.: Windows of vulnerability: a case study analysis. Computer **33**(12), 52–59 (2000)
2. Caplan, S.E.: Problematic internet use and psychosocial well-being: development of a theory-based cognitive-behavioral measurement instrument. Comput. Hum. Behav. **18**(5), 553–575 (2002)
3. Caplan, S.E.: Preference for online social interaction: a theory of problematic internet use and psychosocial well-being. Commun. Res. **30**(6), 625–648 (2003)
4. Cohen, S., Mermelstein, R., Kamarck, T., Hoberman, H.M.: Measuring the functional components of social support. In: Social support: Theory, research and applications, pp. 73–94. Springer (1985)
5. Davis, R.A.: A cognitive-behavioral model of pathological internet use. Comput. Hum. Behav. **17**(2), 187–195 (2001)
6. Gałuszka, A., Probierz, E.: Problematyczne używanie internetu a cechy osobowości i wczesne nieadaptacyjne schematy użytkowników sieci. Annales Universitatis Paedagogicae Cracoviensis. Studia de Cultura **1**(10 (4)" Cyberpsychologia. Nowe strategie badania mediów i ich użytkowników"), 40–50 (2018)
7. Gosling, S.D., Rentfrow, P.J., Swann Jr., W.B.: A very brief measure of the big-five personality domains. J. Res. Pers. **37**(6), 504–528 (2003)

8. Hillard, D., Purpura, S., Wilkerson, J.: Computer-assisted topic classification for mixed-methods social science research. J. Inf. Technol. Polit. **4**(4), 31–46 (2008)
9. Kim, K.S., Kim, K.H.: A prediction model for internet game addiction in adolescents: using a decision tree analysis. J. Korean Acad. Nurs. **40**(3), 378–388 (2010)
10. Lazega, E., Van Duijn, M.: Position in formal structure, personal characteristics and choices of advisors in a law firm: a logistic regression model for dyadic network data. Soc. Netw. **19**(4), 375–397 (1997)
11. Malarvizhi, R., Thanamani, A.S.: K-nearest neighbor in missing data imputation. Int. J. Eng. Res. Dev. **5**(1), 5–7 (2012)
12. Oei, T.P., Baranoff, J.: Young schema questionnaire: review of psychometric and measurement issues. Aust. J. Psychol. **59**(2), 78–86 (2007)
13. Pontes, H.M., Caplan, S.E., Griffiths, M.D.: Psychometric validation of the generalized problematic internet use scale 2 in a portuguese sample. Comput. Hum. Behav. **63**, 823–833 (2016)
14. Probierz, E.: Problematyczne Używanie Internetu a cechy osobowoóci, wczesne nieadaptacyjne schematy, wsparcie społeczne i samoocena użytkowników sieci. (Problematic Internet Use in the context of personality features, early maladaptive schemas, social support and self-esteem of network users.). Master's thesis, Univeristy of Silesia (2018)
15. Robins, R.W., Hendin, H.M., Trzesniewski, K.H.: Measuring global self-esteem: construct validation of a single-item measure and the rosenberg self-esteem scale. Pers. Soc. Psychol. Bull. **27**(2), 151–161 (2001)

FIT2COMIn – Robust Clustering Algorithm for Incomplete Data

Krzysztof Siminski[(✉)]

Institute of Informatics, Silesian University of Technology,
ul. Akademicka 16, 44-100 Gliwice, Poland
krzysztof.siminski@polsl.pl

Abstract. In the paper we propose a new fuzzy interval type-2 C-ordered-means clustering algorithm for incomplete data. The algorithm uses both marginalisation and imputation to handle missing values. Thanks to imputation values in incomplete items are not lost, thanks to marginalisation imputed data can be distinguished from original complete items. The algorithm elaborates rough fuzzy sets (interval type-2 fuzzy sets) to model imprecision and incompleteness of data. For handling outliers the algorithm uses loss functions, ordering technique, and typicalities. Outliers are assigned with low values of typicalities. The paper describes also a new imputation technique–imputation with values from k nearest neighbours.

Keywords: Clustering · Outliers · Noise · Missing values ·
Marginalisation · Imputation · Loss functions ·
Interval type-2 fuzzy sets

1 Introduction

Clustering is one of the essential techniques of unsupervised data analysis. It can reveal patterns in presented data. Real life data often have various faults as outliers or missing values that cause problems in data analysis and can even completely distort the results of clustering.

Many techniques have been proposed for clustering of noisy data. The main approaches are: modification of objective function in clustering (eg. possibilistic clustering [6], repulsive clusters [20], alternative C-means clustering [22], clustering with weaker dependency on parameters [24]), application of special metrics, belief function [9]; data ordering technique [23] and typicalities [7,8].

Data may suffer from incompleteness. It may be caused by errors in data acquisition, random noise, impossible values, merging of data from various sources. The incomplete data may hold important information and its analysis is a challenging task. Generally three approaches are used to handle the problem of missing values: (1) marginalisation of incomplete data items [21] or incomplete data attributes [1], (2) imputation of missing values [12,17], and (3) application of rough set to model incompleteness [5,11,15,16]. Marginalisation is the

© Springer Nature Switzerland AG 2020
A. Gruca et al. (Eds.): ICMMI 2019, AISC 1061, pp. 99–110, 2020.
https://doi.org/10.1007/978-3-030-31964-9_10

simplest approach. It leaves only complete data in a data set. Marginalisation of incomplete data items reduces size of the data set. When all data item are incomplete, marginalisation yields an empty data set. Marginalisation of incomplete attributes leads to reduction of dimensionality of the task domain. Important attributes may be deleted. Although the imputation is more complicated than marginalisation, it is used more frequently. Many techniques of imputation have been proposed (imputation with constants, averages, medians or sophisticatedly elaborated values). The comparison of imputation techniques for clustering of missing data is provided in [10]. The essential problem with the imputation is the fact that imputed values cannot be fully trusted, they may have no physical meaning in the data set, or have values not existing in real life. The only technique of those enumerated above that keeps distinction between original and imputed data is application of rough sets [15].

In the paper we propose a new imputation technique–imputation with values from k nearest neighbours, and a novel fuzzy interval type-2 C-ordered-means clustering algorithm for incomplete data. Our algorithm is an extension of fuzzy C-ordered means clustering algorithm [8] and rough fuzzy clustering [15]. The algorithm elaborates rough fuzzy clusters (interval type-2 fuzzy sets, IT2) to model imprecision (with primary membership functions) and incompleteness of data (with secondary membership functions), and ordering technique with loss functions to make the algorithm robust to outliers and noise.

In the paper we follow the general rule for symbols: the blackboard bold uppercase characters ($\mathbb{A}$) are used to denote sets, uppercase italics (A) – cardinality of sets, uppercase bolds ($\mathbf{A}$) – matrices, lowercase bolds ($\mathbf{a}$) – vectors, lowercase italics (a) – scalars and set elements.

2   Fuzzy Interval Type-2 C-Ordered Means for Incomplete Data

Fuzzy Interval Type-2 C-Ordered Means for Incomplete data (FIT2COMIn) clustering algorithm is designed for clustering incomplete data. It applies both marginalisation and imputation. Imputation saves information in incomplete partial data (lost in marginalisation) and marginalisation keeps a subset of complete data and enables distinguishing imputed complete data from original complete data. The algorithm applies rough fuzzy sets (interval type-2 fuzzy sets) to handle incompleteness of data and ordering technique with loss functions and data typicalities to make the algorithm robust to outliers and noise. This technique can successfully detect outliers.

---

1   **procedure** FIT2COMIn ($\mathbb{X}$, C, m, $\mathcal{L}$)
2   // $\mathbb{X}$: set of data items, C : number of clusters, $m = 2$: weighting
            exponent for memberships, $\mathcal{L}$: loss function
3   ($\underline{\mathbb{X}}$, $\mathbb{X} \setminus \underline{\mathbb{X}}$) ← split ($\mathbb{X}$); // split $\mathbb{X}$ into complete and incomplete
            data sets
4   ($\mathbb{I}, \mathbf{w}$) ← impute ($\mathbb{X}$) ; // $\mathbb{I}$: set of imputed data items, $\mathbf{w}$: weights
            of imputed data items

5    $\overline{\mathbb{X}} \leftarrow \underline{\mathbb{X}} \cup \mathbb{I}$;

6   // final clustering:

7    $(\mathbf{V}, \overline{\mathbf{T}}, \underline{\mathbf{T}}, \overline{\mathbf{S}}, \underline{\mathbf{S}}) \leftarrow \text{FIT2COM}(\underline{\mathbb{X}}, \overline{\mathbb{X}})$;   // $\mathbf{V}$: centres of clusters, $\overline{\mathbf{S}}$:
     "upper" fuzziness of clusters, $\underline{\mathbf{S}}$: "lower" fuzziness of clusters, $\overline{\mathbf{T}}$:
     typicalities of "upper" data items, $\underline{\mathbf{T}}$: typicalities of "lower" data
     items

8   // reconstruct typicalities for incomplete data:

9    $\mathbf{g} \leftarrow \text{reconstruct\_typicalities} (\overline{\mathbf{T}}, \underline{\mathbf{T}})$;

10   **return** $(\mathbf{V}, \mathbf{g}, \overline{\mathbf{S}}, \underline{\mathbf{S}})$;

11 **end procedure**;

Fuzzy Interval Type-2 C-Ordered Means for Incomplete data (FIT2COMIn) clustering algorithm has four steps: data split, imputation of missing values, Fuzzy Interval Type-2 C-Ordered Means clustering, and reconstruction of typicalities for incomplete data.

Data Split. First the data set $\mathbb{X}$ is split into two sets: a set of complete data items $\underline{\mathbb{X}}$ and a set of incomplete data items $\mathbb{X} \setminus \underline{\mathbb{X}}$. Then incomplete data are imputed. The imputed data items and complete items build "upper" data set $\overline{\mathbb{X}}$. In general $\underline{\mathbb{X}} \subseteq \overline{\mathbb{X}}$. When the data set is complete, then $\mathbb{X} = \underline{\mathbb{X}} = \overline{\mathbb{X}}$. In our approach the "lower" data set $\underline{\mathbb{X}}$ keeps the original information not distorted by imputation. "Upper" data set $\overline{\mathbb{X}}$ keeps information lost in marginalisation. For clustering an extension of the fuzzy C-ordered mean (FCOM) clustering algorithm [8] is used. Two data sets $\underline{\mathbb{X}}$ and $\overline{\mathbb{X}}$ are simultaneously clustered so that interval type-2 fuzzy clusters are elaborated. When the data set is complete, interval type 2 fuzzy sets become type 1 fuzzy sets. In the paper we use the following convention: Overlined symbols $\overline{o}$ denote values elaborated with "upper" data set $\overline{\mathbb{X}}$. Similarly underlined symbols $\underline{o}$ stand for values elaborated with "lower" data set $\underline{\mathbb{X}}$. These are not a pair of rough values.

Imputation. Any technique of imputation can be used in the clustering algorithm. In experiments we use imputation (without regard for the cause of missing values) with: averages of missing attributes, medians of missing attributes, averages of k nearest neighbours ($k = 3, 5, 10$), medians of k nearest neighbours ($k = 3, 5, 10$), imputation with values of k nearest neighbours ($k = 3, 5, 10$). The last technique—imputation with values of k nearest neighbours—a new approach is described below. From all data tuples $\mathbb{X}$ (both complete and incomplete) k nearest neighbours are selected. The neighbours may be incomplete, but they must have values required for imputation (other attributes may be missing). The distance $\langle \mathbf{x}_i, \mathbf{x}_j \rangle$ between two incomplete data items $\mathbf{x}_i = [x_{i1}, x_{i2}, \ldots, x_{iD}]$ and $\mathbf{x}_j = [x_{j1}, x_{j2}, \ldots, x_{jD}]$ with D attributes is calculated with formula $\langle \mathbf{x}_i, \mathbf{x}_j \rangle = \frac{\sum_{d=1}^{D} \Delta(x_{id}, x_{jd})}{\sum_{d=1}^{D} [x_{id}, x_{jd}]}$, where the numerator represents a sum of distances between attributes in both tuples $\Delta(x_{id}, x_{jd}) = (x_{id} - x_{jd})^2$ if both values x_{id} and x_{jd} exist, and 0 otherwise. The denominator is a number of existing common attributes in both tuples: $[x_{id}, x_{jd}] = 1$, if both values x_{id} and x_{jd} exist, and 0 otherwise [2]. The neighbours must have attributes needed for imputation. If a data tuple has not a needed attribute, it is not regarded as a neighbour. Each incomplete data item is substituted with exactly k complete data items. The missing values of an incomplete data item are imputed with values from the first neighbour and the first imputed data item is composed. The ith imputed data item inherits missing values from the ith nearest neighbour. This technique prevents from a burst in a number of imputed data items, when missing values are imputed with all combinations of values from neighbours [14]. Because imputed data items are not as reliable as original data, their weight is lower. One incomplete data item is substituted with k data item whose weights are $w = \frac{1}{k}$ each. The imputed data items are not used to impute other incomplete data items. Only original data are used to impute incomplete data items. We decided to use this imputation technique to avoid large increase in amount of imputed data, especially when a data item lacks several attributes. Some algorithms apply imputation of m missing attributes with all combinations of n values. This results in m^n extra data items and makes this technique unpractical in real life applications [19].

Clustering. The Fuzzy Interval Type-2 C-Ordered Means (FIT2COM) clustering algorithm is based on minimisation of a criterion function J:

$$J\left(\overline{\mathbf{U}}, \underline{\mathbf{U}}, \mathbf{V}, \overline{\mathbf{T}}, \underline{\mathbf{T}}\right) = \sum_{c=1}^{C} \left[\sum_{i=1}^{\overline{X}} \overline{t}_{ci} \left(\overline{u}_{ci}\right)^m \|\mathbf{v}_c, \overline{\mathbf{x}}_i\| + \sum_{i=1}^{X} \underline{t}_{ci} \left(\underline{u}_{ci}\right)^m \|\mathbf{v}_c, \underline{\mathbf{x}}_i\| \right], \quad (1)$$

where (we only explain overlined symbols, underlined symbols have analogous meanings) $\overline{\mathbf{U}} = \{\overline{u}_{ci}\}$ is a matrix of memberships (of the ith data item $\overline{\mathbf{x}}_i$ from $\overline{\mathbb{X}}$ to the cth cluster), $\mathbf{V} = [\mathbf{v}_1, \ldots, \mathbf{v}_C]^{\mathrm{T}}$ holds cluster centres, $\overline{\mathbf{T}} = \{\overline{t}_{ci}\}$ is a matrix of typicalities (of the ith data item from $\overline{\mathbb{X}}$ to the cth cluster); C stands for a number of clusters, $\overline{X}$ – number of data items in $\overline{\mathbb{X}}$, D – number of attributes (descriptors), and $\|\mathbf{v}_c, \overline{\mathbf{x}}_i\| = \sum_{d=1}^{D} \mathcal{L}\left(v_{cd} - \overline{x}_{id}\right)$ is a distance of $\overline{\mathbf{x}}_i$ data item from c-th cluster centre $\mathbf{v}_c$. We use following loss functions:

- linear: $\mathcal{L}_l(x) = |x|$;
- Huber, $\delta > 0$ (in experiments: $\delta = 0.3$): $\mathcal{L}_h(x) = \begin{cases} \frac{x^2}{\delta^2}, & |x| \leqslant \delta \\ \frac{|x|}{\delta}, & |x| > \delta \end{cases}$;
- sigmoidal (with parameters: $\alpha, \beta > 0$, in experiments: $\alpha = 1$, $\beta = 1$): $\mathcal{L}_s(x) = \frac{1}{1+\exp[-\alpha(|x|-\beta)]}$;
- sigmoidal linear (with parameters: $\alpha, \beta > 0$, in experiments: $\alpha = 1$, $\beta = 1$): $\mathcal{L}_{sl}(x) = \frac{|x|}{1+\exp[-\alpha(|x|-\beta)]}$;
- logarithmic: $\mathcal{L}_{log}(x) = \log\left[1 + x^2\right]$;
- logarithmic linear $\mathcal{L}_{logl}(x) = |x| \log\left[1 + x^2\right]$.

For minimising of the criterion function J (cf Eq. (1)) with Lagrange multipliers following constrains have been defined:

$$\forall_{k\in\underline{\mathbb{X}}}\quad \sum_{c=1}^{C}\underline{t}_{ck}\underline{u}_{ck}=\underline{f}_{k}, \qquad \forall_{k\in\overline{\mathbb{X}}}\quad \sum_{c=1}^{C}\overline{t}_{ck}\overline{u}_{ck}=\overline{f}_{k}w_{k}, \tag{2}$$

where $\underline{t}_{ck}$ (respectively $\overline{t}_{ck}$) is a typicality of the kth data item $\underline{\mathbf{x}}_{k}$ ($\overline{\mathbf{x}}_{k}$) to the cth cluster, $\underline{f}_{k}$ (respectively $\overline{f}_{k}$) is a global typicality of kth item in $\underline{\mathbb{X}}$ (respectively $\overline{\mathbb{X}}$) data set (Eq. 5), and w_{k} is a weight of the kth data item in $\overline{\mathbb{X}}$ (if the data item has been imputed its lower is less then 1). The Lagrangian G of (1) with constraints is used to minimise the function and yields:

$$\overline{u}_{sk} = \frac{\overline{f}_{k}w_{k}\left\|\mathbf{v}_{s},\overline{\mathbf{x}}_{k}\right\|^{\frac{1}{1-m}}}{\sum_{c=1}^{C}\overline{t}_{ck}\left\|\mathbf{v}_{c},\overline{\mathbf{x}}_{k}\right\|^{\frac{1}{1-m}}}, \qquad \underline{u}_{sk} = \frac{\underline{f}_{k}\left\|\mathbf{v}_{s},\underline{\mathbf{x}}_{k}\right\|^{\frac{1}{1-m}}}{\sum_{c=1}^{C}\underline{t}_{ck}\left\|\mathbf{v}_{c},\underline{\mathbf{x}}_{k}\right\|^{\frac{1}{1-m}}}. \tag{3}$$

```
1  procedure  calculate_cluster_centres  (X)
2      V ← initialize cluster centres with 0's;
3      foreach cluster c
4          do  // begin loop
5              foreach attribute d
6                  foreach data item x_k
7                      calculate residuals r_kdc = |x_kd − v_cd|
8                  end foreach;
9                  rank the residuals form the lowest to the largest;
10                 foreach data item x_k
11                     t_ckd ← elaborate typicality for each ranked residual
                           with PLOWA, SOWA, or UOWA;
12                 end foreach;
13                 update the prototype for the d-th attribute with Eq (4);
14             end foreach;
15             calculate typicality t_ck of kth item to cth cluster;
16          until ‖V^(i) − V^(i+1)‖_F < ε;
17      end foreach;
18      return (V, T);   // cluster centres and typicality
19  end procedure;
```

Calculation of Cluster Centres and Typicalities.

The centres of clusters can be calculated either with "lower" $\underline{\mathbb{X}}$ or "upper" $\overline{\mathbb{X}}$ data items

$$\mathbf{v}_{cd} = \frac{\sum_{k=1}^{\overline{X}}\overline{u}_{ck}^{m}\mathcal{L}\left(\overline{\mathbf{x}}_{d}-\mathbf{v}_{cd}\right)w_{k}\overline{t}_{k}\mathbf{x}_{kd}}{\sum_{k=1}^{\overline{X}}\overline{u}_{ck}^{m}\mathcal{L}\left(\overline{\mathbf{x}}_{d}-\mathbf{v}_{cd}\right)w_{k}\overline{t}_{k}} \quad \text{or} \quad \mathbf{v}_{cd} = \frac{\sum_{k=1}^{\underline{X}}\underline{u}_{ck}^{m}\mathcal{L}\left(\underline{\mathbf{x}}_{d}-\mathbf{v}_{cd}\right)\underline{t}_{k}\mathbf{x}_{kd}}{\sum_{k=1}^{\underline{X}}\underline{u}_{ck}^{m}\mathcal{L}\left(\underline{\mathbf{x}}_{d}-\mathbf{v}_{cd}\right)\underline{t}_{k}}. \tag{4}$$

In experiments we use the first (left) approach. Because $\mathbf{v}_{cd}$ are on both sides of equations and because of application of loss functions $\mathcal{L}$ the localisations of clusters are elaborated with an iterative algorithm. This procedure elaborates both localisations of cluster centres $\mathbf{V}$ and typicalities of data items to cluster $\mathbf{T}$.

Calculation of cluster centres needs an explanation. For each attribute of each data item the typicality is calculated with residuals. A residual r_{kdc} of the k-th data item from c-th cluster with regard to the d-th attribute is defined as $r_{kdc} = |\mathbf{x}_{kd} - \mathbf{v}_{cd}|$. Then the residuals are sorted in nondescending order and each residual is labelled with a number l from 1 (the closest residual) to X (the farthest residual). Then each residual is assigned with typicality t calculated with function $t = \max\left[\min\left(\frac{cX-l}{2bX} + \frac{1}{2}\right); 0\right]$ called piecewise linearly ordered weighted averaging (PLOWA) or $t = \left(1 + \exp\left[\frac{2.944}{aX}\left(l - cX\right)\right]\right)^{-1}$ called sigmoidally ordered weighted averaging (SOWA) [8]. In both functions l stands for the index of the kth data item after ranking by the distance from the c-th cluster with respect to the d-th attribute. One more function is defined: uniform ordered weighted averaging (UOWA), this function has the same value for all ls: $t = 1$.

Typicality t_{ck} of the k-th data item to the c-th cluster is calculated as a T-norm (product) of typicalities of all its attributes $t_{ck} = \prod_{d=1}^{D} t_{ckd}$. The global typicality $\underline{f}_i$ (and respectively $\overline{f}_i$) of the i-th data item is an S-norm $\diamond$ (maximum) of typicalities $\underline{t}_{ci}$ ($\overline{t}_{ci}$) of this data item to all clusters:

$$\underline{f}_i = \underline{t}_{1i} \diamond \underline{t}_{2i} \diamond \ldots \diamond \underline{t}_{Ci}, \qquad \overline{f}_i = \overline{t}_{1i} \diamond \overline{t}_{2i} \diamond \ldots \diamond \overline{t}_{Ci}. \qquad (5)$$

One step in the calculation of cluster centres needs an explanation. In initial iterations of the algorithm the typicality mechanism is switched off. It prevents from incorrect localisation of cluster centres. The experiments show that the number of iteration without typicalities should be small (approximately 5–10).

Upper $\overline{s}$ and lower $\underline{s}$ fuzzification of d-th descriptor in c-th cluster are calculated as

$$\overline{s}_{cd} = \sqrt{\frac{\sum_{i=1}^{\overline{X}} \overline{t}_{ci}\left(\overline{u}_{ci}\right)^m \mathcal{L}\left(\overline{x}_{id} - v_{cd}\right)}{\sum_{i=1}^{\overline{X}} \overline{t}_{ci}\left(\overline{u}_{ci}\right)^m}}, \qquad \underline{s}_{cd} = \sqrt{\frac{\sum_{i=1}^{X} \underline{t}_{ci}\left(\underline{u}_{ci}\right)^m \mathcal{L}\left(\underline{x}_{id} - v_{cd}\right)}{\sum_{i=1}^{X} \underline{t}_{ci}\left(\underline{u}_{ci}\right)^m}}. \qquad (6)$$

```
1  procedure FIT2COM
2      initialise U̅, U̲ with random values with constraints (2);
3      do // begin loop
4          update U̅, U̲ with Eq. (3);
5          (V̅, T̅) ← calculate_cluster_centres (X̅, w̅);
6          // or if you wish
7          // (V̲, T̲) ← calculate_cluster_centres (X̲);
8          calculate global typicalities with (5);
9      until ‖V^(i) − V^(i+1)‖_F < ε;
10     calculate S̅, S̲ with Eq. (6);
11 end procedure;
```

The final step of FIT2COMIn algorithm is reconstruction of typicalities of incomplete data. For a complete data item $\mathbf{x}$ its final typicality is a S-norm of 'upper' $\bar{t}(\mathbf{x})$ and 'lower' $\underline{t}(\mathbf{x})$ typicalities (in our implementation: maximum). For an incomplete data item $\mathbf{x}$ the final typicality is calculated as an S-norm of typicalities of all imputed data items that origin from the incomplete data item $\mathbf{x}$ in question.

3 Experiments

3.1 Datasets

In description of data sets the data items are of two types: informative (regular) data items, non informative data items (outliers).

The data set **'art-outliers'** is a two attribute data set with data items in 3 clusters with 100 items each localised at $(1,5)$, $(5,1)$, $(5,5)$ with Gaussian distribution $\sigma = 1$. The outliers at $(-1,-1)$ with Gaussian distribution $\sigma = 0.1$ are added to the clusters described above. The number of outliers is 0, 10, 20, 30, 40, and 50. In each data set the ratio of missing values is 0, 1, 2, 5, 10, 20, and 50%. The data items were generated $n = 13$ times for each pair: number of outliers and ratio of missing values. This data set is similar to data set used in [3, 8, 18].

The **'Concrete'** set is a real life data set describing the parameters of a concrete sample and its strength [4, 25]. For our experiments normalised data sets with 0%, 1%, 2%, 5%, 10%, 20%, and 50% of outliers and 0%, 1%, 2%, 5%, 10%, 20%, and 50% of values missing have been prepared.

The data set **'Methane'** contains the real life measurements of air parameters in a coal mine in Upper Silesia (Poland) [13]. The data has 1022 data items. The values of attributes in the data set are normalised to interval $[0,1]$. The original data set is accompanied with data set with added 1%, 2%, 5%, 10%, 20%, and 50% outliers in point $[-0.5,\ldots,-0.5]$ and 1%, 2%, 5%, 10%, 20%, and 50% missing values (at random).

3.2 Numerical Experiments

Artificial Data Set with Varying Number of Outliers: 'art-outliers'. The cluster centres in 'art-outliers' data sets are known. The results elaborated by the clustering algorithm are evaluated with the Frobenius norm $E = \|\mathbf{V}_e - \mathbf{V}_0\|_F$, where $\mathbf{V}_0 = \begin{bmatrix} -5 & -5 \\ -5 & -5 \\ -5 & -5 \end{bmatrix}$ is a matrix of real centres of clusters; $\mathbf{V}_e$ is a matrix of elaborated cluster centres. The Frobenius norm of a matrix $\mathbf{A}$ is defined as $\|\mathbf{A}\|_F = \sum_{i,j} a_{ij}^2$. When the cluster centres match perfectly, $E = 0$.

The experiments for the 'art-outliers' data set were conducted for various combinations of parameters.

We tested the proposed clustering algorithm for various: ($\star$) weighting functions (SOWA, PLOWA, and UOWA); ($\star\star$) loss functions (linear, Huber, sigmoidal [with parameters: $\alpha = 1$, $\beta = 1$], sigmoidal linear [with parameters: $\alpha = 1$, $\beta = 1$], logarithmic, logarithmic linear), and ($\star\star\star$) imputation methods (imputation with averages, median, averages of $k = 3, 5, 10$ nearest neighbours, medians of $k = 3, 5, 10$ nearest neighbours, imputation with values from $k = 3, 5, 10$ nearest neighbours).

It is impossible to show all elaborated results in the article (6 outliers numbers $\times 7$ missing ratios $\times 3$ OWAs $\times 6$ loss functions $\times 11$ imputation techniques $\times 13$ repetitions $= 108108$ experiments) so we present here only the most important ones.

In figure captions we use following abbreviations for loss functions: lin – linear, hub – Huber, log – logarithmic, lln – logarithmic linear, sig – sigmoidal, slg – sigmoidal linear; and for imputation methods: avg – imputation with averaged, med – with medians, 3nna – average of 3 nearest neighbours, 5nna – average of 5NN, 3nnm – median of 3NN.

Figure 1 presents the results elaborated for all tested loss functions. Because some of them result in very high values of Frobenius norm, Fig. 2 shows only a subset of those that have lower values of the norm. The lowest values of Frobenius norm are elaborated with the logarithmic linear loss function for number of outliers not greater than 30. Good news is the fact that for outlier number up to 30 the logarithmic linear function keeps the error on the same level. However for the number of outliers greater than 30 the more robust loss function is the sigmoidal one (Fig. 1). Three weighting function were tested in experiments: UOWA, SOWA, and PLOWA. The results are presented in Fig. 3. Although all functions do not manage ratio 50 of outliers in the data set, for lower values of outliers the most efficient one is SOWA.

For lower ratios of missing values in the data set the most efficient imputation methods are based on medians (Fig. 4). However it is worth mentioning that for higher ratios of missing values these methods of imputation cause wrong identification of cluster locations. In this case it is better to use imputation with values from k-nn values (Fig. 5).

Figure 6 presents a histogram of typicalities assigned to data items in 'art-outliers' data set with 40 outliers. The outliers are assigned with the lowest typicalities represented by the most left bin in the histogram.

'Concrete' – Concrete Compressive Strength. A huge number of experiments with 'Concrete' data set have been run to test possible combinations of imputation methods, loss functions, and OWA functions. Here again the number of experiments is huge (7 outliers numbers $\times 7$ missing ratios $\times 3$ OWAs $\times 6$ loss functions $\times 11$ imputation techniques $\times 13$ repetitions $= 126126$ experiments) and it is impossible to present all their results in the paper.

In case of 'Concrete' data set the real number of clusters is not known. This is why we estimated the number by clustering a data set with no extra outliers added and no missing values with fuzzy c-mean (FCM) algorithm and then testing the elaborated partition with Xie-Beni clustering index $\nu_{XB}(\mathbf{U}) =$

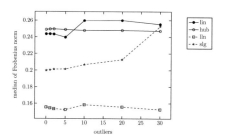

Fig. 1. Medians of Frobenius norm for a SOWA and no missing values.

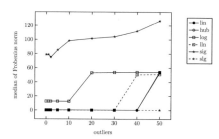

Fig. 2. Medians of Frobenius norm for a SOWA and no missing values.

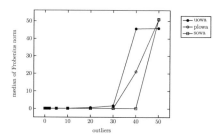

Fig. 3. Medians of Frobenius norm for a logarithmic-linear dissimilarity function and no missing values and imputation with values from knn ($k = 10$).

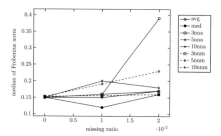

Fig. 4. Medians of Frobenius norm for a logarithmic-linear dissimilarity function, outliers ratio 2, and various imputations techniques.

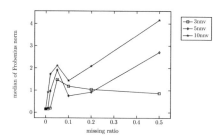

Fig. 5. Medians of Frobenius norm for a logarithmic-linear dissimilarity function, outliers ratio 2, and imputation with values of $k = 3, 5, 10$ neighbours.

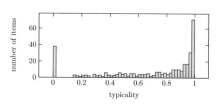

Fig. 6. Histogram of typicalities of data item elaborated with logarithmic linear dissimilarity function and SOWA for 'art-outliers' data set with 40 outliers. A bin located at the left for near zero typicalities represents outliers.

$$\frac{\sum_{c=1}^{C} \sum_{k=1}^{X} (u_{ck})^m \|\mathbf{v}_c - \mathbf{x}_k\|^2}{X(\min_{i \neq j} \|\mathbf{v}_i - \mathbf{v}_j\|^2)},$$ where X stands for number of data items, $\mathbf{x}_k$ – kth data item, $\mathbf{v}_c$ – centre of cth cluster. The best (lowest) value of ν_{XB} is achieved for $C = 3$ clusters. Then the FIT2COMIn algorithm was run for $C = 3$ for various ratios of outliers and ratios of missing values.

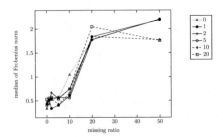

Fig. 7. Frobenius norm for imputation with medians of $k = 3$ nearest neighbours for various ratio of outliers in 'Concrete' data set.

Fig. 8. Frobenius norm for imputation with values of $k = 3$ nearest neighbours for various ratio of outliers in 'Concrete' data set.

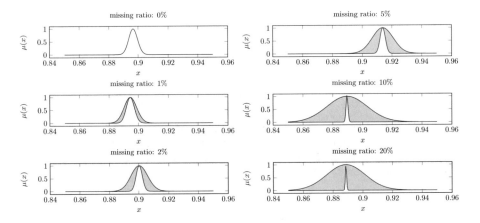

Fig. 9. Interval type-2 fuzzy sets elaborated for the first attribute of 'Methane' data set with no outliers and various ratios of missing values. The grey area represents the footprint of uncertainty of interval type-2 fuzzy sets.

For 'Concrete' data set the most efficient is the logarithmic linear loss function with sigmoidally ordered weighted averaging (SOWA). Figures 7 and 8 present medians of Frobenius norm for various number of outliers and ratios of missing values elaborated with the FIT2COMIn algorithm with $k = 3$ nearest neighbours median imputation and imputation with values from $k = 3$ nearest neighbours respectively. Imputation with medians of $k$nn is the most stable one – the influence of outliers is the weakest. In case of median imputation the higher ratio of outliers (20) deteriorates the elaborated results. Imputation with values from $k$nn is the best method for missing ratio 20 and outlier ratios 1–10. Other imputation methods elaborates poorer results.

'Methane' – Methane Concentration. The experiments were run for the 'Methane' data set for the same combination of parameter as for 'Concrete' data set. The number of cluster was estimated $C = 5$ with the Xie-Beni index as for the 'Concrete' data set (Sect. 3.2). The results are similar to those for the 'Concrete'. This is why we do not present them here. We would like to present interval type-2 fuzzy sets elaborated for attributes. Figure 9 presents fuzzy sets for the first attribute of 'Methane' data set for no outliers and various ratios of missing values. For a data set with no missing values interval type-2 fuzzy sets degenerate into type-1 fuzzy sets. For data sets with missing values the greater missing value ratio, the greater the footprint of uncertainty in clusters. The elaborated clusters represent both imprecision of data and their uncertainty.

4 Conclusions

The paper presents a fuzzy c-ordered clustering algorithm for incomplete data (FIT2COMIn). The algorithm has been designed to handle incomplete data with outliers. The algorithm uses both marginalisation and imputation to handle incomplete data. In experiments eleven imputation techniques have been used (among them a proposed new imputation algorithm). Any other imputation technique may be used. The algorithm elaborates clusters built with interval type-2 fuzzy sets, that model both imprecision and uncertainty of incomplete data. The algorithm uses non-Euclidean distance measures (five loss functions) to better handle outliers. The algorithm assigns data items with typicalities. Non-informative data items (outliers) are assigned with low values of typicalities. The ordering technique with three weighted averaging functions have been used in experiments. The experiments show that the proposed algorithm can handle a number of outliers almost as high as a number of informative items in a cluster. The most promising loss function from all researched is the logarithmic loss function with sigmoidal ordered weighted averaging function.

Acknowledgements. The research has been supported by the Rector's Grant for Research and Development (Silesian University of Technology, grant number: 02/020/RGJ19/0165).

References

1. Cooke, M., Green, P., Josifovski, L., Vizinho, A.: Robust automatic speech recognition with missing and unreliable acoustic data. Speech Commun. **34**, 267–285 (2001)
2. Dixon, J.K.: Pattern recognition with partly missing data. IEEE Trans. Syst. Man Cybern. **SMC-9**, 617–621 (1979)
3. D'Urso, P., Leski, J.M.: Fuzzy clustering of fuzzy data based on robust loss functions and ordered weighted averaging. Fuzzy Sets Syst. (2019)
4. Frank, A., Asuncion, A.: UCI machine learning repository (2010)
5. Grzymała-Busse, J.: A rough set approach to data with missing attribute values. In: Wang, G., Peters, J., Skowron, A., Yao, Y. (eds.) Rough Sets and Knowledge Technology. Lecture Notes in Computer Science, vol. 4062, pp. 58–67. Springer, Heidelberg (2006)

6. Krishnapuram, R., Keller, J.: A possibilistic approach to clustering. IEEE Trans. Fuzzy Syst. **1**, 98–110 (1993)
7. Leski, J., Kotas, M.: On robust fuzzy c-regression models. Fuzzy Sets Syst. **279**, 112–129 (2015)
8. Leski, J.M.: Fuzzy c-ordered-means clustering. Fuzzy Sets Syst. **286**, 114–133 (2014)
9. Masson, M.-H., Denœux, T.: ECM: an evidential version of the fuzzy c-means algorithm. Pattern Recogn. **41**, 1384–1397 (2008)
10. Matyja, A., Siminski, K.: Comparison of algorithms for clustering incomplete data. Found. Comput. Decis. Sci. **39**(2), 107–127 (2014)
11. Nowicki, R.: Rough-neuro-fuzzy system with MICOG defuzzification. In: 2006 IEEE International Conference on Fuzzy Systems, Vancouver, Canada, pp. 1958–1965 (2006)
12. Renz, C. Rajapakse, J.C., Razvi, K., Liang, S.K.C.: Ovarian cancer classification with missing data. In: Proceedings of the 9th International Conference on Neural Information Processing, ICONIP 2002, Singapore, vol. 2, pp. 809–813 (2002)
13. Sikora, M., Sikora, B.: Application of machine learning for prediction a methane concentration in a coal-mine. Arch. Min. Sci. **51**(4), 475–492 (2006)
14. Siminski, K.: Neuro-rough-fuzzy approach for regression modelling from missing data. Int. J. Appl. Math. Comput. Sci. **22**(2), 461–476 (2012)
15. Siminski, K.: Clustering with missing values. Fundamenta Informaticae **123**(3), 331–350 (2013)
16. Siminski, K.: Rough subspace neuro-fuzzy system. Fuzzy Sets Syst. **269**, 30–46 (2015)
17. Siminski, K.: Imputation of missing values by inversion of fuzzy neuro-system. In: Gruca, A., Brachman, A., Kozielski, S., Czachórski, T. (eds.) Man–Machine Interactions 4, pp. 573–582. Springer, Cham (2016)
18. Siminski, K.: Fuzzy weighted c-ordered means clustering algorithm. Fuzzy Sets Syst. **318**, 1–33 (2017)
19. Siminski, K.: NFL - free library for fuzzy and neuro-fuzzy systems. In: Kozielski, S. (ed.) Beyond Databases, Architectures and Structures. Paving the Road to Smart Data Processing and Analysis, pp. 139–150. Springer, Cham (2019)
20. Timm, H., Borgelt, C., Döring, C., Kruse, R.: An extension to possibilistic fuzzy cluster analysis. Fuzzy Sets Syst. **147**, 3–16 (2004)
21. Troyanskaya, O., Cantor, M., Sherlock, G., Brown, P., Hastie, T., Tibshirani, R., Botstein, D., Altman, R.B.: Missing value estimation methods for DNA microarrays. Bioinformatics **17**(6), 520–525 (2001)
22. Kuo-Lung, W., Yang, M.-S.: Alternative c-means clustering algorithms. Pattern Recogn. **35**, 2267–2278 (2002)
23. Yager, R.R.: On ordered weighted averaging aggregation operators in multicriteria decisionmaking. IEEE Trans. Syst. Man Cybern. **18**(1), 183–190 (1988)
24. Yang, M.-S., Kuo-Lung, W.: Unsupervised possibilistic clustering. Pattern Recogn. **39**, 5–21 (2006)
25. Cheng Yeh, I.: Modeling of strength of high-performance concrete using artificial neural networks. Cem. Concr. Res. **28**(12), 1797–1808 (1998)

GrFCM – Granular Clustering
of Granular Data

Krzysztof Siminski[✉]

Institute of Informatics, Silesian University of Technology,
ul. Akademicka 16, 44-100 Gliwice, Poland
krzysztof.siminski@polsl.pl

Abstract. Granular computing is a new paradigm in data mining. It
mimics a procedure commonly used by humans. A data granule may be
defined as a collection of related entities in sense of similarity, proximity,
indiscernibility. Nowadays granular computing focuses on elaboration of
granules from data. This step in granular computing is well researched.
Our objective is the next step: we would like to focus on computing with
granules. In the paper we propose a new clustering algorithm that works
with granules instead of numbers. The algorithm takes a collection of
granules as an input and clusters them into output granules.

Keywords: Clustering · Granule · Granular computing

1 Introduction

1.1 Data Granules

Granular computing is an emerging field of research in data analysis. A data
granule is a composed entity with clear semantics. This new paradigm is a shift
from computer-centred to human-centred analysis. Human's view on the outer
world is based on granulation. We trace granular entities in the environment and
then compose hierarchical structures of granules.

The phrase *information granulation* was coined by Lotfi Zadeh in 1979 [16].
However this idea was not exploited until 1997 when Zadeh published a paper
in which he described the idea of «granularcomputing»[17]. He stated three
concepts of human cognition: granulation (decomposition of whole into parts),
organization (integration of parts into whole), and causation (relations of causes
and effects). Granular computing is a machine implementation of this human
cognition model. In recent years granular computing has waken up from a long
hibernation and has made a huge progress [11]. Yao [13] claims granular computing a new field of study.

A data granule may be defined as a collection of related entities in sense of
similarity, proximity, indiscernibility [10,12,14]. A granule represents a semantic whole and simultaneously is in a hierarchical relation with other granules: a

© Springer Nature Switzerland AG 2020
A. Gruca et al. (Eds.): ICMMI 2019, AISC 1061, pp. 111–121, 2020.
https://doi.org/10.1007/978-3-030-31964-9_11

granule is a component of a more general granule and simultaneously is a composition of subcomponents–subgranules. The granulation process enables smooth transition from detailed to general view of data. The view on the data can be easily zoomed in and out [4]. Granules are commonly represented with fuzzy sets [15], rough sets [8], shadowed sets [9], fuzzy rough sets [2], intuitionistic fuzzy sets [1], etc.

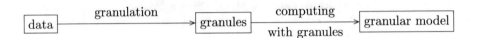

Fig. 1. A scheme of granular data processing.

Nowadays granular computing focuses on value granulation of data–elaboration of granules from the presented data. Common used techniques are: discretization, quantization, clustering, aggregation, transformation. This step in granular computing is well researched, many papers on data granulation have been published. Our objective is the next step in the granular data processing (Fig. 1). We want to focus on computing with granules. This requires analysis, proposals, and implementation of algorithm working with granules instead of numbers.

1.2   Extensional Fuzzy Numbers

In our approach we use granules defined with extensional fuzzy numbers [5–7]. This is a very interesting construction of numbers because it addresses an important problem in usage of fuzzy numbers. Application of Zadeh's extension principle in construction of fuzzy numbers has two big practical drawbacks: (1) The more operations are executed on fuzzy numbers, the more fuzzy the result is – often even beyond practical application; (2) The results of operations on fuzzy numbers not always have (or even often do not have) a form of a fuzzy number. This is why we use a new construction of fuzzy numbers proposed by Holčapek and Štěpnička [5–7]. The numbers are constructed in such a way that they both keep their fuzziness and the results of operations are still legal fuzzy numbers – they are almost an algebraic field what makes their application possible. It is *almost a field* because multiple neutral elements are defined in their algebraic structure. It may seem strange, but it is very intuitive: we may define several fuzzy zeros (*exactly zero, almost zero, more or less zero*, etc). Zadeh's extension principle requires that a field of fuzzy number has one zero: a crisp zero. This is mathematically correct, but quite counterintuitive: in a field of *fuzzy* numbers zero is *crisp*.

In our approach we construct Gaussian extensional fuzzy numbers with a drastic T-norm and similarity relation defined as $S_\sigma(x,y) = \exp\left(-\frac{(x,y)^2}{2\sigma^2}\right)$. In this approach a fuzzy number p is represented by a pair $p = \langle \mathfrak{C}(p), \mathfrak{F}(p) \rangle$, where

$\mathfrak{C}(p)$ stands for a core of a fuzzy number p and $\mathfrak{F}(p)$ for its fuzziness. Essential operations are defined as:

$$\langle\mathfrak{C}(p),\mathfrak{F}(p)\rangle + \langle\mathfrak{C}(q),\mathfrak{F}(q)\rangle = \langle\mathfrak{C}(p) + \mathfrak{C}(q),\max\{\mathfrak{F}(p),\mathfrak{F}(q)\}\rangle \quad (1)$$
$$\langle\mathfrak{C}(p),\mathfrak{F}(p)\rangle \cdot \langle\mathfrak{C}(q),\mathfrak{F}(q)\rangle = \langle\mathfrak{C}(p) \cdot \mathfrak{C}(q),\max\{\mathfrak{F}(p),\mathfrak{F}(q)\}\rangle \quad (2)$$

For the brevity of the paper we do not describe the algebra of extensional fuzzy numbers. For details please consult original papers [5–7].

1.3   Granular Fuzzy C-Means Clustering Algorithm

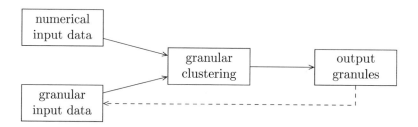

Fig. 2. Data flow of a granular clustering.

Our aim is to modify a clustering algorithm (and in further future other data mining algorithms) to work both on numbers and granules (Fig. 2). We start with the widely known FCM algorithm. It has its drawbacks (eg. sensibility to outliers), but is simple and commonly known in the community. In Fig. 3 we present a *granular fuzzy c-means* (GrFCM) algorithm in detail. Input data are denoted as a collection of input granules $\mathbb{G} = \{\mathbf{g}_1,\ldots,\mathbf{g}_n\}$. Each granule $\mathbf{g}_i = [g_{i1},\ldots,g_{ia}]$ is a collection of extensional fuzzy numbers representing descriptors of a granule. In the GrFCM the result of clustering is a collection $\Gamma = [\gamma_{i1},\ldots,\gamma_{ca}]$ of granules. The form of an input $\mathbf{g}$ and an output γ granule is exactly the same: a collection of extensional fuzzy numbers representing descriptors – we just use Latin symbol for input and Greek for output granules to distinguish them. The algorithm differs from the fuzzy c-means clustering algorithm in the form of elaborated result. FCM elaborates fuzzy membership for each data item to each cluster. Our algorithm elaborates granules – fuzzy points representing fuzzy prototypes.

2   Experiments

The objective of the experiments is to show that the proposed algorithm is able to elaborate (very) similar results to those returned by the well known FCM algorithm. Time complexity of FCM is $O(CXDI)$, where C stands for a number of clusters, X – number of data items in a data set, D – number of attributes, and I – number of iterations. FCM has linear complexity with regard to number of items to cluster. Thus splitting a data set into subsets is pointless. But very big data that cannot be stored in the memory and the memory is a bottleneck of the algorithm. The data cannot be clustered as a whole and have to be clustered by parts. In experiments we want to test the granular clustering algorithm and then show its application in a *per partes* hierarchical clustering of large data sets.

---

1 **procedure** GrFCM $(\mathbb{G},\ c)$ *// granular FCM*
2   *// $\mathbb{G} = \{\mathbf{g}_1, \ldots, \mathbf{g}_n\}$ – set of input data (granules)*
3   *// c – number of output granules (clusters) to elaborate*
4      $\mathbf{U}' = \begin{bmatrix} u'_{11} & \cdots & u'_{1n} \\ \vdots & \ddots & \vdots \\ u'_{c1} & \cdots & u'_{cn} \end{bmatrix} \leftarrow \text{randomise}$;   *// partition matrix*
5   **do**
6          $\mathbf{U} \leftarrow \mathbf{U}'$;
7   **foreach** c **do**   *// for each cluster (output granule)*
8              $\boldsymbol{\gamma}_c \leftarrow \frac{\sum_{i=1}^{n} u_{ci}^m \mathbf{g}_i}{\sum_{i=1}^{n} u_{ci}^m}$;   *// calculate its centre (granule core)*
9   **end foreach** ;
10   *// calculate distances of each data item to each cluster centre:*
11   **foreach** c **do**   *// for each cluster (output granule)*
12   **foreach** i **do**   *// for each data item*
13                  $d_{ci}^2 \leftarrow \|\boldsymbol{\gamma}_c - \mathbf{g}_i\|^2$;   *// Gaussian distance*
14                  $D_c \leftarrow D_c + d_{ci}^2$;
15   **end foreach** ;
16   *// modify partition matrix:*
17   **foreach** i **do**   *// for each data item*
18                  $u'_{ci} \leftarrow \frac{d_{ci}^2}{D_c}$;
19   **end foreach** ;
20   **end foreach** ;
21   **while** $(\|\mathbf{U} - \mathbf{U}'\|_F > \varepsilon)$;   *// Frobenius norm*
22
23   *// elaborate fuzzification of granules:*
24      $\Gamma \leftarrow \varnothing$; *// empty set of output granules*
25   **foreach** c **do**   *// for each cluster (output granule)*
26   **foreach** a **do**   *// for each attribute elaborate its fuzziness*
27              $\mathfrak{F}(\gamma_{ca}) \leftarrow \mathfrak{F}(\gamma_{ca}) + \mathfrak{C}\left(\sqrt{\frac{\sum_{i=1}^{n} u_{ci}^m (g_{ia} - \gamma_{ca})^2}{\sum_{i=1}^{n} u_{ci}^m}} \right)$;
28   **end foreach** ;
29          $\Gamma \leftarrow \Gamma \cup \{\boldsymbol{\gamma}_c\}$;
30   **end foreach** ;
31   **return** $\Gamma = \{\gamma_1, \gamma_2, \ldots, \gamma_c\}$;   *// return elaborated granules*
32 **end procedure** ;

Fig. 3. Pseudocode of fuzzy c-means algorithm.

In clustering of data blocks we use a hierarchical partition paradigm. Original data (level 0 data) are clustered by parts. Elaborated granules build level 1 data. In general: if there are enough level i data (more than l_i), they are clustered into c_i granules that are added to level $(i+1)$ data. Then the next block of n_0 original (level 0) data are read in and clustered. And again if necessary the granules are clustered and results added to level 1 data. The scheme of the hierarchical paradigm is presented in Fig. 4 and the pseudocode in Fig. 5.

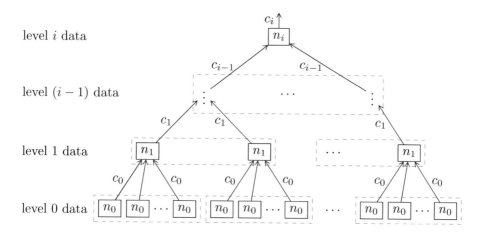

Fig. 4. A scheme of a hierarchical paradigm.

2.1 Data Sets

In experiments we use

- two artificial two dimensional data sets with four clusters located at $(1,1)$, $(1,5)$, $(5,1)$, $(5,5)$ with fuzzification of each attribute $\sigma = 1$ and data numbers
  - 40 000 for the *small* data set and
  - 10 000 000 for the *large* data set.
- a data set with overlapping data clusters located at $(1,1)$, $(1,5)$, $(5,1)$, $(5,5)$ with fuzzification of each attribute $\sigma = 3$ and 10 000 000 data items (*overlapping* data set).
- benchmark BIRCH data sets with 100 000 vectors and 100 clusters [3,18]:
  - clusters in regular grid structure (*birch-1* dataset),
  - clusters at a sine curve (*birch-2* dataset).

```
1  procedure hierarchical_paradigm;
2      level_data : array of data granules;
3      while exist data to read do
4          i ← 1;
5          data ← read a part input data;
6          granules ← GrFCM (data, c_i);
7          level_data[i] ← level_data[i] ∪ granules;
8
9          while (||level_data[i]|| > l_i) then
10             granules ← GrFCM (level_data[i], c_i);
11             level_data[i+1] ← level_data[i+1] ∪ granules;
12             level_data[i] ← ∅;
13             i ← i + 1;
14         end while;
15     end while;
16     // All data are read in.
17     i ← 1;
18     while (level_data[i] ≠ ∅) do
19         if (||level_data[i]|| > t_i) then
20             granules ← GrFCM (level_data[i], c_i);
21             level_data[i+1] ← level_data[i+1] ∪ granules;
22         else
23             level_data[i+1] ← level_data[i+1] ∪ level_data[i];
24         end if;
25         i ← i + 1;
26     end while;
27     return level_data[i−1]; // final granules
28 end procedure;
```

Fig. 5. Pseudocode of a hierarchical paradigm.

2.2   Results

First we discuss experiments conducted for the *small* (with 40 000 items) and *large* data set (with 10 000 000 items). The former one enables direct execution of FCM algorithm.

Number n_0 of original data items read in one block are different for *small* and *large* datasets. For the *small* data set (with 40 000 data items) we use $n_0 \in \{5, 10, 20, 50, 100, 200, 500, 1\,000, 2\,000, 5\,000, 10\,000, 20\,000\}$, for the *large* data set (with 10 000 000 data items) $n_0 \in \{10, 20, 50, 100, 200, 500, 1\,000, 2\,000, 5\,000, 10\,000, 20\,000, 50\,000, 100\,000, 200\,000, 500\,000, 1\,000\,000\}$. In experiments we use parameters: $c_i = 4$, $t_i = 10c_i$, and $l_i = \frac{n_0}{2}$ (Figs. 4 and 5).

The aim is to test if our method can elaborate clusters as FCM. In all experiments clusters (output granules) have been correctly localized with quite good precision, for example a typical elaborated localisation of output granules: $(0.967199, 0.967392)$, $(0.967218, 5.03362)$, $(5.03212, 0.967633)$, $(5.03265, 5.03134)$. The fuzzification of elaborated granules differ from those elaborated by FCM.

The results for the *small* data set are presented in Fig. 6 and Table 1. For this data set it is possible to run FCM directly. The fuzzification of output granules is larger than those elaborated by FCM, but asymptomatically tends to values elaborated by FCM with the increase of data block size n_0. It is interesting that execution time depends on data block size. For very small blocks execution time is quite long. However for small blocks $n_0 = 5, \ldots, 50$ it takes shorter time to cluster data by parts than with FCM. For medium and large blocks execution time grows with data block size. This may be explained by two facts: (1) The size of blocks influences the number of levels (as defined in Fig. 4) in the clustering algorithm. For very small blocks number of level is greater than 10, for medium and large blocks ($n_0 > 100$) there are only two levels in data clustering. The most interesting is the number of levels for n_0 that results in the shortest execution time (number of levels is 4–5). (2) For big data blocks the extra time overhead is caused by memory operations.

For the *large* data set it is not possible to run FCM directly because of memory shortage. But the pattern can be seen again, although we do not have the results of FCM for this data set. For very small data blocks ($n_0 < 100$) both fuzzification and execution time are large. The minimum of execution time is for small data block $n_0 = 200$ (data set size $N = 10^7$). Here again the number of levels follows the pattern elaborated for the *small* data set.

The size of blocks is essential for the quality of granules. It has weak influence on localisations of output granules, but heavily influences the fuzzification of granules and execution time. In order to get similar results as FCM algorithm blocks of data should be as large as possible (small blocks result in wider fuzzification), but for large blocks memory is a bottleneck that lengthens execution time. A reasonable trade-off has to be set: several (4–5) levels in the hierarchical clustering approach (Fig. 7).

The similar conclusions may be drawn for the *overlapping* data set with 10 000 000 items. It is impossible to run experiments with the whole data set read in as a whole (as in FCM) due to lack of memory. The results are presented in Table 1.

Several experiments were also run on the benchmark datesets *birch-1* (Figs. 8 and 9) and *birch-2* (Figs. 10 and 11). The experiments were run for $n_0 = 10000, 20000, 50000, 100000, 200000, 500000, 1000000$. The last one $n_0 = 1000000$ means that the whole data set was read in one part. The scheme of results is similar to the previous datasets: For smaller values of n_0 the fuzziness of elaborated granules is higher. For the *birch-1* data set the a slightly higher concentration of output granules in the centre of the data set is visible (Fig. 8). This phenomenon is not noticeable for the *birch-2* data set (Fig. 10).

Table 1. Fuzzyfication of elaborated clusters and clustering time for the *small* and *large* data sets.

n_0	*small* data set		*large* data set		*overlapping* data set	
	Fuzzyfication	Time [ms]	Fuzzyfication	Time [s]	Fuz-tion	Time [s]
5	2.91 ± 0.09	879 ± 72				
10	2.91 ± 0.09	393 ± 56	21.29 ± 1.63	548 ± 31		
20	2.51 ± 0.11	332 ± 42	6.20 ± 0.07	410 ± 31		
50	1.92 ± 0.04	426 ± 76	2.89 ± 0.02	345 ± 15		
100	1.57 ± 0.02	629 ± 52	2.21 ± 0.02	220 ± 19		
200	1.35 ± 0.00	717 ± 87	1.69 ± 0.12	209 ± 16		
500	1.20 ± 0.00	795 ± 87	1.32 ± 0.00	209 ± 16		
1 000	1.15 ± 0.00	846 ± 93	1.21 ± 0.00	216 ± 19		
2 000	1.06 ± 0.00	854 ± 57	1.10 ± 0.00	228 ± 26		
5 000	1.01 ± 0.00	935 ± 81	1.04 ± 0.00	237 ± 31		
10 000	0.99 ± 0.00	969 ± 59	1.01 ± 0.00	250 ± 30	2.73 ± 0.00	4065 ± 198
20 000	0.97 ± 0.00	1008 ± 64	1.00 ± 0.00	262 ± 28	2.56 ± 0.00	4687 ± 134
50 000			0.98 ± 0.00	285 ± 38	2.39 ± 0.00	4858 ± 167
100 000			0.97 ± 0.00	270 ± 4	2.32 ± 0.00	5052 ± 304
200 000			0.97 ± 0.00	280 ± 4	2.27 ± 0.00	5236 ± 456
500 000			0.96 ± 0.00	292 ± 6	2.23 ± 0.00	5266 ± 577
1 000 000			0.96 ± 0.00	299 ± 5	2.21 ± 0.00	5409 ± 564
2 000 000			0.96 ± 0.00	293 ± 6	2.20 ± 0.00	5736 ± 1105
5 000 000			0.95 ± 0.00	305 ± 5	2.18 ± 0.00	5831 ± 1183
FCM	0.95 ± 0.00	535 ± 60				

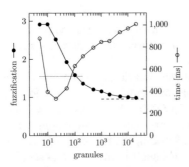

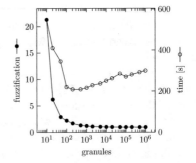

Fig. 6. Fuzzyfication of elaborated clusters and clustering time for the *small* data set. The dashed line represents fuzzyfication elaborated by FCM, dotted line execution time of FCM algorithm.

Fig. 7. Fuzzyfication of elaborated clusters and clustering time for the *large* data set.

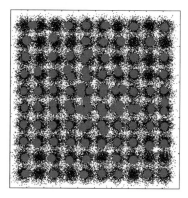

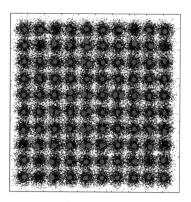

Fig. 8. Localisation of points (black dots) in the *birch-1* data set and localisation of cluster (red ellipses) elaborated with GrFCM with $n_0 = 10000$.

Fig. 9. Localisation of points (black dots) in the *birch-1* data set and localisation of cluster (red ellipses) elaborated with GrFCM with $n_0 = 1000000$, what is equivalent to FCM clustering.

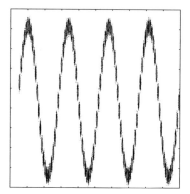

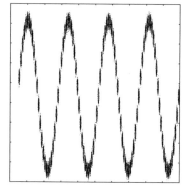

Fig. 10. Localisation of points (black dots) in the *birch-2* data set and localisation of cluster (red ellipses) elaborated with GrFCM with $n_0 = 10000$.

Fig. 11. Localisation of points (black dots) in the *birch-2* data set and localisation of cluster (red ellipses) elaborated with GrFCM with $n_0 = 1000000$, what is equivalent to FCM clustering.

3   Conclusions

In the paper we describe *granule fuzzy c-means clustering algorithm* (GrFCM) that implements FCM algorithm for granules. It works both on numbers and granules. The algorithm elaborates similar results to the base (FCM) clustering algorithm. In the paper we have proposed an application of the GrFCM algorithm in a *per partes* hierarchical clustering paradigm for data that cannot be clustered as a whole. The blocks of numerical data are clustered into granules. If the set of granules is large enough it is further clustered. This procedures continues until the whole data set is clustered.

Acknowledgements. The research has been supported by the Rector's Grant for Research and Development (Silesian University of Technology, grant number: 02/020/RGJ19/0165).

References

1. Atanassov, K.: Intuitionistic fuzzy sets. Seventh Scientific Session of ITKR (1983)
2. Dubois, D., Prade, H.: Rough fuzzy sets and fuzzy rough sets. Int. J. Gen Syst **17**(2), 191–209 (1990)
3. Fränti, P., Sieranoja, S.: K-means properties on six clustering benchmarkdatasets. Appl. Intell. **48**(12), 4743–4759 (2018). https://doi.org/10.1007/s10489-018-1238-7. http://cs.uef.fi/sipu/datasets/
4. Grzegorowski, M., Janusz, A., Slezak, D., Szczuka, M.: On the role of feature space granulation in feature selection processes. In: 2017 IEEE International Conference on Big Data (Big Data), pp. 1806–1815, December 2017. https://doi.org/10.1109/BigData.2017.8258124
5. Holčapek, M., Štěpnička, M.: Arithmetics of extensional fuzzy numbers – part I: Introduction. In: 2012 IEEE International Conference on Fuzzy Systems, pp. 1–8 (2012). https://doi.org/10.1109/FUZZ-IEEE.2012.6251274
6. Holčapek, M., Štěpnička, M.: Arithmetics of extensional fuzzy numbers – part II: Algebraic framework. In: 2012 IEEE International Conference on Fuzzy Systems, pp. 1–8 (2012). https://doi.org/10.1109/FUZZ-IEEE.2012.6251275
7. Holčapek, M., Štěpnička, M.: MI-algebras: a new framework for arithmetics of (extensional) fuzzy numbers. Fuzzy Sets Syst. **257**, 102–131 (2014). https://doi.org/10.1016/j.fss.2014.02.016. Special Issue on Fuzzy Numbers and Their Applications. http://www.sciencedirect.com/science/article/pii/S0165011414000682
8. Pawlak, Z.: Rough sets. Int. J. Parallel Prog. **11**(5), 341–356 (1982)
9. Pedrycz, W.: Shadowed sets: representing and processing fuzzy sets. Trans. Sys. Man Cyber. Part B **28**(1), 103–109 (1998). https://doi.org/10.1109/3477.658584
10. Pedrycz, W.: Granular Computing: Analysis and Design of Intelligent Systems. CRC Press, Boca Raton (2013)
11. Yao, J.T., Vasilakos, A.V., Pedrycz, W.: Granular computing: perspectives and challenges. IEEE Trans. Cybern. **43**(6), 1977–1989 (2013). https://doi.org/10.1109/TSMCC.2012.2236648
12. Yao, Y., Zhong, N.: Granular computing. In: Wiley Encyclopedia of Computer Science and Engineering (2008)
13. Yao, Y.: The art of granular computing. In: Proceeding of the International Conference on Rough Sets and Emerging Intelligent Systems Paradigms LNAI, vol. 4585, pp. 101–112 (2007)
14. Yao, Y.: Granular computing: past, present and future. In: The 2008 IEEE International Conference on Granular Computing, GrC 2008, Hangzhou, China, 26–28 August 2008, pp. 80–85 (2008). https://doi.org/10.1109/GRC.2008.4664800
15. Zadeh, L.A.: Fuzzy sets. Inf. Control **8**, 338–353 (1965)
16. Zadeh, L.A.: Fuzzy sets and information granularity. In: Gupta, N., Ragade, R., Yager, R. (eds.) Advances in Fuzzy Set Theory and Applications, pp. 3–18 (1979)

17. Zadeh, L.A.: Toward a theory of fuzzy information granulation and its centrality in human reasoning and fuzzy logic. Fuzzy Sets Syst. **90**, 111–127 (1997). https://doi.org/10.1016/S0165-0114(97)00077-8
18. Zhang, T., Ramakrishnan, R., Livny, M.: BIRCH: a new data clustering algorithm and its applications. Data Mining Knowl. Discov. **1**(2), 141–182 (1997)

Pattern Recognition

Evaluation of Cosine Similarity Feature for Named Entity Recognition on Tweets

Onur Büyüktopaç and Tankut Acarman[✉]

Computer Engineering Department, Galatasaray University, Ortaköy,
34349 İstanbul, Turkey
onurbuyuktopac@gmail.com, tacarman@gsu.edu.tr

Abstract. In this paper, we present the Named Entity Recognition as a multi-class system and we evaluate baseline classifiers along with the technical features extracted from tweet datasets. Initially, we elaborate the conversion procedure of tweet data and we study three different datasets such that raw tweet data are compatible to the presented data model. The first dataset is well-known for benchmarking purposes and the other two datasets have been collected in the wild by using Twitter search API and given keywords. Then, we elaborate the feature vector constituted by 9 technical features. To reach at higher statistical metric values of the multi-class NER system, we seek the performance of the classifier subject to different combination of features. Finally, we elaborate the impact of the cosine similarity to the class centroid feature to the performance of the classifiers and we present the highest F1 score reached by using a particular set of features.

Keywords: Named entity recognition · Information extraction ·
Twitter · .Classification · Cosine similarity

1 Introduction

Named Entity Recognition (NER) identifies and categorizes textual contents such as person, thing, organization, location, product, event and character. While traditional NER approaches classify successfully when working with well structured texts, the prediction scores are much lower on unstructured microblog texts like Twitter. These type of texts contain emoticons, abbreviations, grammar mistakes and mixed languages. In this study, we present an improved approach using syntactic, semantic and domain specific features while augmenting the dataset used in [9]. For this purpose we extend the dataset, use feature scaling and evaluate all feature combinations. For benchmarking purposes, we use publicly available dataset provided by Named Entity rEcognition and Linking (NEEL) 2016 Challenge, [8]. A feature based approach with Word2Vec in addition to syntactic and domain specific features is elaborated in [9]. This study is an extension of [9] while focusing on the most representative feature set analysis

© Springer Nature Switzerland AG 2020
A. Gruca et al. (Eds.): ICMMI 2019, AISC 1061, pp. 125–135, 2020.
https://doi.org/10.1007/978-3-030-31964-9_12

and the improvement of the classifier's performance. In [4] a feature based system combining existing NER systems and domain specific Part-Of-Speech (POS) tagger is presented. Candidate name generation and classical NER systems such as Stanford NER, MITIE, twitter_nlp and TwitIE are used in [6]. An adapted Kanopy system for Twitter domain is implemented in [10].

2 Tweet Data and Features

2.1 Tweet Data

We consume Named Entities (NEs) from 3 different corpora with different layouts. For unifying the layouts, we create a common format. The common format is tab-separated and it stores the "tweet id", "tweet text", "NE starting index", "NE ending index", "NER type" and "NE" information as follows:

[tweet_ID] [text] [NE_starting_index] [NE_ending_index] [NER_type] [NE]

One tweet can contain multiple NEs but each NE is stored individually. For benchmarking purposes, we use publicly available dataset provided by NEEL 2016 Challenge [8]. The source contains only labeled word information and tweet ID. The template of the source file can be described as follows:

[tweet_ID] [word_start_index] [word_end_index] [word_Dbpedia_link] [confidence_score] [NER_type]

For instance:

674869443671941120 93 101 http://dbpedia.org/resource/Egyptians 1 Thing

Since some features require the content of tweet to calculate its score, we need to merge the information above with the tweet. Finding tweets from the tweet ID is a challenging task, which is also elaborated in [9], because either some tweets were already deleted or they are private. Therefore, we cannot retrieve tweets from Twitter. Instead, we extract all tweets from an open source project from GitHub called Web-pack Bundle Analyzer, [7].

The collected tweets are harmonized and merged with the NEEL dataset and stored in a format where each sample contains the following information: tweet identifier, tweet, start index of the word, end index of the word, NER type and the word. The resulting form is illustrated as follows:

674869443671941120 RT @EntheosShines: Just As Some Parents Have A Favorite Child, Obama Has Favorites (sign at Egyptian Airport) @chirofrenzy @PatVPeters htt··· | 93 101 Thing Egyptian

The corpus contains 4369 unique tweets with 9687 labeled words. It consists of 7 different labels which are "Person", "Thing", "Organization", "Location", "Product", "Event", and "Character". The split ratio is 0.10, which is constituted by the training dataset of 4073 unique tweets and 8665 labeled words, and the testing dataset of 296 unique tweets and 1022 labeled words.

The second corpus is provided by a startup company called Oxtractor focussing on social data. The corpus holds several information fields about the tweet including text, ID, retweet count, user profile information, media information, language. However, we only focus on text, ID, entities, language, tokens, and annotation offsets as JSON format. We have simplified the properties for our study and the data structure of a tweet has became compliant to the following structure:

```
{
"text": "Ukraine's pro-Russia rebels hand over Malaysia Airlines
            #MH17's black boxes      http://t.co/sWs4wDau3m
            http://t.co/9GyZCurIkM",
  "id":      491401326845510000,
  "entities":      ["B-loc", "O", "O", "B-loc", "O", "O", "O", "B-org",
                    "I-org", "O", "O", "O", "O", "O"],
  "lang":      "en",
  "tokens":      ["Ukraine", "'s", "pro-", "Russia", "rebels", "hand",
                  "over", "Malaysia", "Airlines", "#MH17's", "black",
                  "boxes", "http://t.co/sWs4wDau3m",
                  "http://t.co/9GyZCurIkM"],
  "annotation_offsets":      [ [0,7], [7,9], [10,14],[14,20], [21,27], [28,32],
                  [33,37], [38,46], [47,55], [56,61], [64,69], [70,75],
                  [76,98], [99,121]]
}
```

In this example, BIO encoding is applied for tokenizing. There are four labeled words and three named entities. "B" key represents the beginning of a named entity, and "I" key represents inside of a named entity. We apply the tab-separated format, then the example transformed into three separate samples as follows:

```
491401326845510000 Ukraine's pro-Russia rebels hand
over Malaysia Airlines #MH17's black boxes
http://t.co/sWs4wDau3m http://t.co/9GyZCurIkM 0 7
   Location Ukraine

491401326845510000 Ukraine's pro-Russia rebels hand
over Malaysia Airlines #MH17's black boxes
http://t.co/sWs4wDau3m http://t.co/9GyZCurIkM 14 20
   Location Russia

491401326845510000 Ukraine's pro-Russia rebels hand
over Malaysia Airlines #MH17's black boxes
http://t.co/sWs4wDau3m http://t.co/9GyZCurIkM 38 55
   Organization Malaysia Airlines
```

The Oxtractor corpus has only 3 label types; "Person", "Organization", and "Location". It contains 4608 unique tweets with 8264 labeled words. We split the data into training and testing by the same ratio used in the previous dataset. Finally, the training dataset consists of 4056 unique tweets with 7395 labeled words, and the testing dataset is constituted by 552 unique tweets with 869 labeled words.

The third corpus has been collected in the wild during 4 months starting July 6th, 2018 and ending October 21st, 2018. We collected English tweets posted in 2018 from Twitter Search API [12]. Keywords cover multiple memorable events mentioned during the collection period such as "World Cup 2018", "U.S. China trade war", "Death of Anthony Bourdain", "Wacken 2018", "Syrian refugee", "Climate Change", "Marvel", "Avengers". While the API provides information about tweet itself, we only consider "ID", "text", and "lang" properties. The basic version of the data is illustrated as follows:

```
{
    "statuses": [
        {
            "id":       1014639601313636352,
            "text":     "4 of 5 stars to Kitchen Confidential
by Anthony Bourdain
            https://t.co/6wpo4qZHIG",
            "lang":     "en",
        }
    ]
}
```

We exclude retweets by a script and we enrich the data with tokenized sentence information:

```
{
            "id":       1014639601313636400,
            "text":     "4 of 5 stars to Kitchen Confidential by
    Anthony Bourdain https://t.co/6wpo4qZHIG",
            "tokens": ["4", "of', "5", "stars", "to", "Kitchen",
        "Confidential","by","Anthony", "Bourdain",
        "https://t.co/6wpo4qZHIG" ]
}
```

Furthermore, we filter the auto-posts from such as news channels and YouTube manually. Then, we label the tweets according to 7 label types, which are "Person", "Thing", "Organization", "Location", "Product", "Event", and "Character". Labeling process was handled manually by using a custom developed tool named Twitter Tagger. The user interface of the tool is presented in Fig. 1.

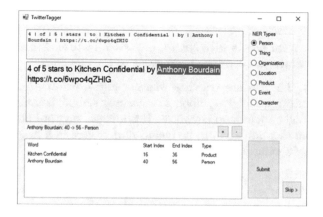

Fig. 1. Twitter Tagger tool interface

After labeling, the raw data is transformed into a tab-separated format, which is compatible to our data model:

1014639601313636400 4 of 5 stars to Kitchen Confidential
by Anthony Bourdain https://t.co/6wpo4qZHIG
16 36 Product Kitchen Confidential

1014639601313636400 4 of 5 stars to Kitchen Confidential by
by Anthony Bourdain https://t.co/6wpo4qZHIG
40 56 Person Anthony Bourdain

NER type occurrences in the three datasets are summarized in Table 1.

Table 1. NER type occurrences in training and testing datasets

Class	NEEL 2016 challenge		Oxtractor		Manually labeled	
	Training	Testing	Training	Testing	Training	Testing
Person	2845	337	3124	531	3338	366
Location	1868	43	2120	122	1023	123
Organization	1641	158	2151	216	1196	154
Product	1196	354	–	—	458	45
Event	482	24	–	–	779	70
Thing	570	49	–	–	644	65
Character	63	57	–	–	67	11
Total	8665	1022	7395	869	7505	834

2.2   Feature Set

We label words from tweets with NER types. Each feature represents a numeric value in the feature vector, which is an input to the classifier.

Hashtag (#). The feature searches whether any word of a named entity is hashtagged, written with "#" symbol. The hashtag symbol is used on Twitter to index and highlight topics and keywords. Hashtagged keyword does not contain any space or punctuation and can be included anywhere in a tweet. The feature returns a Boolean value.

At (@). The feature has the same process with hashtag but it seeks 'at' sign, denoted by "@" symbol, instead of "#". At sign is used for addressing another Twitter user. Twitter usernames can only contain alphanumeric characters and underscore which guarantees mentioned user belongs in NE or not. It returns a Boolean value.

Capital Letter (title). The feature checks whether all words in named entity start with an upper-case letter and the rest of them are lower-case. Symbols and numbers are ignored. It returns a Boolean value.

All Capital (all_capital). The feature checks whether all letters of the named entity are upper-case. Symbols and numbers are also ignored. It returns a Boolean value.

Part-of-Speech (POS) Tagger (pos). The feature assigns a part-of-speech tag to each word of NE such as "noun", "verb", "adjective". We used Stanford POS tagger [11] for this task. We tagged each word of NE separately and checked whether all words have the same tag. If they all have the same POS tag, then NE is assigned with the tag. Otherwise, the NE is labeled as "mixed POS". To apply the results to the feature vector, we map the tags with numeric values.

Next Word POS Tagger (next_pos). The feature assigns a POS tag the following word after NE. If NE is the last word then the feature returns "0".

Previous Word POS Tagger (prev_pos). The feature assigns a POS tag the previous word of NE. If NE is the first word then feature returns "0".

Position (position). The feature presents the position of NE in a tweet. It splits tweet by space and indexes them starting from 1. The first word of NE is accepted as an index of NE. Comparing indexes from different tweets is inconsistent because the index is depended on tweet length. Hence, we normalize the index and define a relative position as shown in (1).

$$position = \frac{index\ of\ the\ first\ word\ of NE}{Total\ word\ count} \tag{1}$$

The Similarity to the Class Centroid (cos). The feature computes cosine similarity between NE's vector and centroid vector of each NER type. All words present in NE's are defined as a vector using the precomputed word2vec model, the corpus of 400 million tweets [5]. Each NE is represented by a single vector which is computed by the weighted average of all word vectors.

All NE vectors are grouped by NER type to calculate the centroid of the types. In this study, we use 7 NER types. Therefore, the similarity to class centroid consists of 7 hidden features. Each hidden feature returns a numeric value.

3   Tests and Evaluation

3.1   Implementation

We implement our algorithm using Python programming language with "genism" library [3] in order to execute word embedding through "word2vec" and "sklearn" library [2] for feature scaling and classifying trained and tested results. We use the following baseline classification algorithms available in the "scikit-learn" library: Logistic Regression, Support Vector Machines (SVM), k-Nearest Neighbors (k-NN), Gaussian Naive Bayes (Gaussian NB), Bernoulli Naive Bayes (Bernoulli NB), Extra Tree, Decision Tree, Random Forest, and Gradient Boosting. For tokenizing and the Part-of-Speech (POS) tagging, we use "nltk" library [1]. The library contains an adapter of Stanford POS Tagger, which is originally written in Java.

3.2   Test and Results

The NEEL 2016 dataset is used to evaluate the performance of classifiers subject to the combination of features. The SVM classifier with RBF kernel using the feature **"title"**, **"position"**, **"next_pos"**, and **"cos"** achieves the highest statistical metric values such as the precision of 0.74, recall of 0.68 and an F1 micro average of 0.67. The classifier predicts 691 true labels correctly from 921 ground truth labels, the confusion matrix is given in Table 2.

Table 2. Confusion matrix of the SVM classifier with RBF kernel and "cos", "all capital" and "#" features for 7 NER Types

Class	Person	Thing	Organization	Location	Product	Event	Character
Person	321	7	1	4	4	0	0
Thing	69	61	2	18	7	1	0
Organization	7	3	29	2	2	0	0
Location	5	4	0	38	2	0	0
Product	134	8	3	4	203	2	0
Event	10	1	0	1	0	12	0
Character	19	2	0	0	9	0	27

In Table 2, the majority of the misclassified NEs are labeled as "Person" even though most of the actual "Person" NEs are labeled correctly. The prediction performance is high on recall but low on precision. In contrast, the "Character" label has no false positive value but half of the actual values is falsely labeled. "Location" has the highest F1 score with 0.74 and "Organization" has the lowest F1 score with 0.5. Table 3 presents the statistical metric values of the SVM classifier with RBF kernel for each label type or class.

Table 3. The classification report of the features

Class	Precision	Recall	F1 score
Person	0.57	0.95	0.71
Organization	0.71	0.39	0.5
Location	0.83	0.67	0.74
Thing	0.57	0.78	0.66
Product	0.89	0.57	0.7
Event	0.8	0.5	0.62
Character	1	0.47	0.64

In Table 4, the precision, recall, and F1 statistical metric values are benchmarked by using the dataset provided by NEEL 2016 Challenge [8]. The methods in [4,6,10] presented during the NEEL 2016 workshop and the method presented by [9] are benchmarked and compared in Table 4.

Table 4. The performance of our approach is compared with other studies

Study	Precision	Recall	F1 score
Our method (SVM with RBF, 3 features + Cosine Similarity)	0.74	0.68	0.68
A feature based approach performing Stanford NER, [4]	0.73	0.63	0.67
Logistic Regression, 5 features + Cosine Similarity, [9]	0.71	0.56	0.58
TwitIE (CRF Model), [10]	0.44	0.46	0.45
Stanford NER, MITIE, twitter_nlp and TwitIE, [6]	0.59	0.29	0.39

The class centroid (cos) feature has a significant impact to the classifier performance in comparison with respect to the suite of other 8 features. In Table 5, the F1 statistical metric value achieved by classifiers with and without (denoted by 'w/o') the presence of the "cos" feature in the feature suite are compared. The highest F1 metric values assured by 3 classifiers using the feature suite including the "cos" feature are Logistic Regression, SVM and Random Forest. The ranking varies among 3 classifiers depending on the particular dataset. The performance of Logistic Regression and SVM decreases dramatically when the "cos" feature is excluded. The most successful 3 classifiers using the combination of features excluding the "cos" feature are Random Forest, Decision Tree, and k-NN.

Table 5. The impact of the cosine similarity "cos" feature to the F1 metric

Classifier	NEEL 2016 challenge			Oxtractor			Manually labeled		
	With	w/o	Average	With	w/o	Average	With	w/o	Average
Logistic Regression	0.56	0.12	0.34	0.74	0.43	0.58	0.79	0.17	0.48
SVM with RBF	0.57	0.14	0.36	0.72	0.45	0.59	0.78	0.23	0.51
SVM with Linear	0.57	0.11	0.34	0.74	0.42	0.58	0.81	0.18	0.49
k-NN	0.47	0.14	0.31	0.68	0.45	0.56	0.76	0.29	0.53
Gaussian NB	0.34	0.12	0.23	0.47	0.39	0.43	0.57	0.15	0.36
Decision Tree	0.37	0.15	0.26	0.65	0.47	0.56	0.64	0.31	0.47
Random Forest	0.51	0.15	0.33	0.71	0.48	0.59	0.82	0.32	0.57

In following, the highest F1 metric values assured by the classifier and the feature combination are given in Table 6. The highest F1 metric value is reached by using the manual dataset followed by the Oxtractor dataset and NEEL 2016 dataset. Three classifiers using the common features such as "title", and "cos" reach the highest F1 metric value.

Table 6. The highest F1 scores for each dataset

Dataset	Score	Classifer	Feature set
NEEL 2016	0.652	SVM with RBF	"title" , "position", "next_pos", and "cos"
Oxtractor	0.767	Logistic Regression	"@" ,"title", "position" , "prev_pos', and "cos"
Manual	0.847	Random Forest	"@" ,"title" , "all_capital", "next_pos", "prev_pos", "cos"

Random Forest classifier assures the highest F1 metric value with respect to other classifiers using the same dataset ratio of the manual dataset. The highest F1 value is reached at 0.859 by using the 80/20 splitting ratio. For the random forest classifier, the F1 statistical metric values are given with respect to the different ratio in Table 7.

Table 7. The highest statistical metric values according to the split ratio

Splitting ratio	Classifer	Precision	Recall	F1 score
90/10	Random Forest	0.882	0.819	0.847
80/20	Random Forest	0.903	0.824	0.859
70/30	Random Forest	0.862	0.784	0.816
60/40	Random Forest	0.860	0.799	0.826
50/50	Random Forest	0.848	0.767	0.802

4   Conclusions

Because of the unstructured nature of the tweets, the features, which are representative on more structured texts like newspapers or articles, do not have an impact to the classifiers. To discover the most representative feature set, we investigate and analyze syntactic, semantic and domain specific features extracted from a richer dataset. First, we describe the descriptive features and analyze the impact of the cosine similarity feature to the NER performance. We seek the highest statistical metric value by combining the 8 features along with the cosine similarity. Overall, we elaborate the performance of the NER system in terms of the features and the ratio of training versus testing dataset.

Acknowledgements. The authors gratefully acknowledge the support of Galatasaray University, scientific research support program under grant #19.401.003.

References

1. Bird, S., Loper, E., Klein, E.: A platform for building Python programs to work with human language data. https://www.nltk.org/. Accessed 8 Mar 2019
2. BlueSoft: A Python module for machine learning. http://sklearn.org/stable/index. html. Accessed 8 Mar 2019
3. Foundation, P.S.: A Python library for topic modelling, document indexing and similarity retrieval with large corpora. https://pypi.python.org/pypi/gensim. Accessed 8 Mar 2019
4. Ghosh, S., Maitra, P., Das, D.: Feature based approach to named entity recognition and linking for tweets. In: #Microposts (2016)
5. Godin, F., Vandersmissen, B., De Neve, W., Van de Walle, R.: Multimedia lab $@$ ACL WNUT NER shared task: named entity recognition for twitter microposts using distributed word representations. In: Proceedings of the Workshop on Noisy User-generated Text, pp. 146–153. Association for Computational Linguistics (2015). https://doi.org/10.18653/v1/W15-4322, http://aclweb. org/anthology/W15-4322
6. Greenfield, K., Caceres, R.S., Coury, M., Geyer, K., Gwon, Y., Matterer, J., Mensch, A.C., Sahin, C.S., Simek, O.: A reverse approach to named entity extraction and linking in microposts. In: #Microposts (2016)
7. Grunin, Y., Laakso, V.: Webpack Bundle Analyzer (2016). https://github.com/ webpack-contrib/webpack-bundle-analyzer. Accessed 8 Mar 2019
8. Rizzo, G., van Erp, M., Plu, J., Troncy, R.: Making sense of microposts (#microposts2016) named entity recognition and linking (neel) challenge. In: Proceedings of the 6th Workshop on 'Making Sense of Microposts', pp. 50–59 (2016)
9. Taşpınar, M., Ganiz, M.C., Acarman, T.: A feature based simple machine learning approach with word embeddings to named entity recognition on tweets. In: Frasincar, F., Ittoo, A., Nguyen, L.M., Métais, E. (eds.) Natural Language Processing and Information Systems, pp. 254–259. Springer International Publishing, Cham (2017)
10. Torres-Tramón, P., Hromic, H., Walsh, B., Heravi, B.R., Hayes, C.: Kanopy4tweets: entity extraction and linking for twitter. In: #Microposts (2016)

11. Toutanova, K., Klein, D., Manning, C.D., Singer, Y.: Feature-richpart-of-speech tagging with a cyclic dependency network. In: Proceedings of the 2003 Conference of the North American Chapter of the Association for Computational Linguistics on Human Language Technology- Volume 1, NAACL 2003, pp. 173–180. Association for Computational Linguistics, Stroudsburg (2003). https://doi.org/10.3115/1073445.1073478
12. Twitter: Standard Search API for Tweets. https://developer.twitter.com/en/docs/tweets/search/api-reference/get-search-tweets.html. Accessed 8 Mar 2019

Deep Recurrent Neural Networks for Human Activity Recognition During Skiing

Magdalena Pawlyta[1,2(✉)], Marek Hermansa[1], Agnieszka Szczęsna[1],
Mateusz Janiak[2], and Konrad Wojciechowski[2]

[1] Institute of Informatics, Silesian University of Technology, Gliwice, Poland
{Magdalena.Pawlyta,Agnieszka.Szczesna}@polsl.pl,
mareher231@student.polsl.pl
[2] Polish-Japanese Academy of Information Technology, Warsaw, Poland
{mpawlyta,mjaniak,konrad.wojciechowski}@pjwstk.edu.pl

Abstract. In recent years, deep learning has been successfully applied to an increasing number of research areas. One of those areas is human activity recognition. Most published papers focus on a comparison of different deep learning models, using publicly available benchmark datasets. This article focuses on identifying specific activity—skiing activity. For this purpose, a database containing information from the three inertial body sensors, placed on skier's chest and on both skis, was created. This database contains synchronized data, from an accelerometer, gyroscope or barometer. Then, two deep models based on the Long Short-Term Memory units, were created and compared. First is a unidirectional neural network which can remember information from the past, second is a bidirectional neural network, which can memorize information from both the past and the future. Both models were tested for different window sizes and the number of hidden layers and the number of units on the layer. These models can be used in the alpine skiing and biathlon training support system.

Keywords: Activity recognition · Deep learning · Recurrent neural networks

1 Introduction

Human activity recognition (HAR) is an active research area having a plurality of applications ranging from surveillance systems, healthcare, personal fitness, gaming to tactical military applications, and human-computer interaction. Research in this domain can be divided into two groups according to the chosen data source: based on sensor [11,16] or based on video data [1,15]. In video-HAR systems, the classification is based on the extraction of a global feature set from a whole video sequence [2,17], or the extraction of a local feature set for one frame or a small set of frames [7,12]. In contrast, sensor-HAR systems, are

© Springer Nature Switzerland AG 2020
A. Gruca et al. (Eds.): ICMMI 2019, AISC 1061, pp. 136–145, 2020.
https://doi.org/10.1007/978-3-030-31964-9_13

based on motion data from different sensors such as accelerometers, gyroscopes, magnetometers, GPS, barometer, etc. In recent research, sensor-HAR systems are becoming more and more popular, due to the thriving sensor development and the growing number of publicly available benchmark datasets [3,4,6,19]. This publication is focused on the recognition of specific activity - skiing activity. Video recording of this type of activity is very problematic due to the high movement dynamics and the large area in which activity is performed. Therefore, the described system will be based on data from inertial sensors.

Generally, sensor based HAR is a time series classification task. Conventional approaches for such type of tasks assume the use of a modified machine learning algorithms like decision tree or support vector machine. However, those algorithms heavily rely on a heuristic, hand-crafted feature extraction. Due to limited human domain knowledge, those feature are usually shallow. Therefore, many state-of-the-art approaches use deep learning models—Convolutional Neural Network (CNN) or a specific type of recurrent neural networks called Long Short-Term Memory Network (LSTM). Initially, CNN was developed for analyzing image data, such as recognition handwritten digits, localizing objects in an image or automatically describing the image content, but they can be successfully adapted for HAR problem [5,8,18]. The second network type, that was designed to learn from sequence data, seems to be more appropriate for sensor-based HAR [10,13].

As it was mentioned, this article focuses on skier activity recognition. For the correct operation of the support alpine skiing and biathlon training system, it is necessary to correctly recognize the following activities:

– left/right turn,
– left/right leg lift,
– ski to ski orientation (plow, parallel or skate),
– body position (race, proper or too much forward, backward or stiff).

Human perception does not allow for immediate reaction to one of the above-mentioned activities—the skier's movement is too dynamic. Therefore, adding a small, few-frame delay will not be noticeable to the user and may improve the effectiveness of the recognition. In order to test this thesis, the use of two LSTM-based models was proposed—uni and bi-directional deep neural network.

2 Sensors Data

Data is collected with Snowcookie® system [14]. It consists of three IMUs (Inertial Measurement Unit) connected to mobile phone (iPhone). Each sensor contains three axis accelerometer, gyroscope and magnetometer. One of those sensors is placed on the chest of the skier, where two others are attached to the skies. Sensors data is sampled with 50 Hz frequency. Additionally, collected data is extended with GPS (position, speed, heading, accuracy) and barometer measurements from the phone, sampled with approximately 1 Hz frequency. All data are converted to SI units (accelerations in m/s^2, angular velocity in rad/s, magnetic flux density in T, pressure in Pa, speed in m/s, heading in rad, position according to WGS84: altitude in m, latitude and longitude in rad).

3   Deep Models

The goal of this section is to provide information about the recurrent neural network (RNN) and Long-Short Term Memory network (LSTM).

3.1   Recurrent Neural Network

An RNN is neural network architecture which enables to memorize some information of past time by using cyclic connections. As shown in Fig. 1, a hidden layer in this type of network contains multiple nodes. Each node calculate the current output o_t and current hidden state h_t using the current input x_t and previous hidden state $h_t - 1$, according to the following formulas:

$$h_t = H(U_x x_t + W_h h_{t-1} + b_h) \tag{1}$$

$$o_t = O(V_h h_t + b_o) \tag{2}$$

where, U_h, W_h, V_h are the connection weights between input and hidden layer, hidden layer and itself, hidden and output layer respectively. b_h, b_o are the bias vector for hidden and output layer. H is the activation functions for hidden and O for an output layer. The commonly chosen activation functions are sigmoid, hyperbolic tangent or rectified linear unit (ReLU).

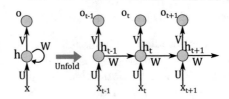

Fig. 1. A recurrent neural network

3.2   Long Short-Term Memory

Nevertheless, RNN has some limitations. Training that type of network is restricted due to the issues of vanishing or exploding gradient. LSTM as an advanced RNN architecture overcomes these issues by replacing nodes with memory block. Those block, shown in Fig. 2, contains special components called gates. Each gate has its own sigmoidal activation function, which controls whether they are active or not. The first gate is the forget gate f_t, which decides what information should be forgotten. Next two gates are responsible for cell memory update.

An input gate i_t decides which value should be updated, and a cell candidate gate C_t create a new vector of value that can be used for that update. The combination of those two gates creates an update of the memory cell. Last is the output gate, which decides what to output. All the activation function, output, and actual memory state can be calculated according to the following formulas:

$$f_t = \sigma(W_f h_{t-1} + U_f x_t + b_f) \tag{3}$$

$$i_t = \sigma(W_i h_{t-1} + U_i x_t + b_i) \tag{4}$$

$$g_t = \sigma(W_g h_{t-1} + U_g x_t + b_g) \tag{5}$$

$$o_t = \sigma(W_o h_{t-1} + U_o x_t + b_o) \tag{6}$$

$$\tilde{C}_t = f_t C_{t-1} + g_t i_t \tag{7}$$

$$h_t = \tanh(C_t) o_t \tag{8}$$

where W, U are weights between each gate and previous hidden state or current input vector respectively, b are the bias vector for each gate, and σ is a sigmoid activation function.

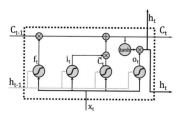

Fig. 2. Schema of the LSTM cell structure

4 Experimental Setup

4.1 Proposed Network Architectures

As it was mentioned before, for this classification problem two models based on LSTM architecture were proposed. Regardless of the type of network chosen, the rest of the model's remains the same—it maps raw multi-modal sensor input to activity label classifications. The inputs signals are segmented into windows of length T and fed into the model. Each window contains a sequence of individual samples observed by the sensor at time t $(x_1, x_2, ..., x_T)$. The proposed models have been tested for different window lengths, the number of hidden layers, and the number of units per layer. The best results were for a window with a length of 32, and three hidden layers with 100 hidden units each. A comparison of these models is presented in the Sect. 5.

Additionally, each hidden layer is followed by a dropout layer to prevent overfitting. In both models, the last layer is a fully connected layer and it is used to perform high-level feature interpretation.

Unidirectional LSTM Recurrent Neural Network Model. The first model was built using a unidirectional LSTM model (uni-LSTM), shown in Fig. 3. In this model, input sequence $(x_1, x_2, ..., x_T)$, are fed into the first layer at time $(t = 1, 2, .., T)$. Value of first hidden state h_0^1 and internal state C_0^1 are set to zeros. The output of each layer is calculated according to Eqs. 3: 8. The outputs of the individual LSTM cells of the last third layer are inputs to the fully connected layer.

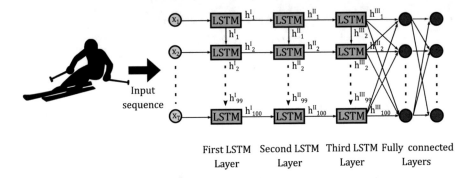

First LSTM   Second LSTM   Third LSTM   Fully connected
   Layer   Layer   Layer   Layers

Fig. 3. The proposed three-layer unidirectional LSTM network. For clarity, the temporal recurrent structure is not shown.

Bidirectional LSTM Recurrent Neural Network Model. The second model was built using bidirectional LSTM model (bi-LSTM), shown in Fig. 4. This model by using two parallel tracks predict the label of each x_1 frame, base on the information from the past and future time step. Like in the unidirectional model value of first hidden state h_0^1 and internal state C_0^1 are set to zeros. The outputs of the individual LSTM cells of the last third layer for both parallel track are inputs to the fully connected layer.

4.2   Network Training

All the models were trained on the same dataset using 72% of the data for training, 18% for validation and 10% for testing. All weights were initialized randomly and then updated to minimize a cost function. The ground truth label for the database was determined by experts who analyzed the partial video material and sensor data.

The Adam stochastic gradient descent [9] was used to optimize the network with initial learning rate set to 0.5 and the categorical cross entropy was used as a loss function since the multi-class problem was considered. Number of epochs was limited by condition of not giving better results in minimizing loss function for 10 consecutive epochs.

Training was conducted on a Keras with use of Tensorflow backend. Keras is high-level neural networks API, written in Python and capable of running

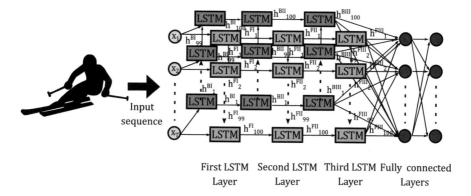

First LSTM   Second LSTM   Third LSTM   Fully connected
   Layer   Layer   Layer   Layers

Fig. 4. The proposed three-layer bidirectional LSTM network. For clarity, the temporal recurrent structure is not shown.

on top of TensorFlow which is an open-source library developed by Google. All models were trained on a NVIDIA Tesla K80.

4.3   Performance Metrics

To measure the performance of the proposed models, four popular metrics for classification were used. F1-score is not differentiable function and as such it can't be used as a loss function. For this reason categorical cross-entropy was used during the training, although F1-score was applied to determine final model effectiveness, since number of classes of some type were unbalanced.

Accuracy. Measure of the proportion of correctly predicted classes over all predictions.

$$Accuracy = \frac{TP + TN}{TP + TN + FP + FN} \tag{9}$$

where TP is the number of correctly identified samples, TN is the number of correctly rejected samples, FP is the number of incorrectly identified samples, and FN is the number of incorrectly rejected samples.

Precision. Measure of the proportion of correctly predicted samples over total predicted positive samples of the class.

$$Precision = \frac{TP}{TP + FP} \tag{10}$$

Recall. Measure of the proportion of correctly predicted samples over total samples of the class.

$$Recall = \frac{TP}{TP + FN} \tag{11}$$

F1-Score. Weighted average of *Precision* and *Recall*.

$$F1\text{-}score = 2 \left(\frac{Recall \cdot Precision}{Recall + Precision} \right) \tag{12}$$

5   Results

Data set used for research contained 420 runs performed by at least 12 different skiers. Aggregated results for accuracy and F1-score shown in Table 1, are similar for both models, however bi-directional model gave slightly better scores in all cases. The most significant difference was for activity—leg lift. Classification 4% better than in the case of a unidirectional model (Fig. 5). Confusion matrices for both models are shown in Figs. 6 and 7.

Table 1. Macro average of F1-score and accuracy of proposed models for each event

		F1-score		Accuracy	
		uni-LSTM	bi-LSTM	uni-LSTM	bi-LSTM
Event	Turn	99.30	99.57	99.33	99.59
	Leg lift	92.55	96.41	99.61	99.78
	ski2ski orientation	99.21	99.43	99.21	99.43
	Body position	99.03	99.10	99.13	99.19

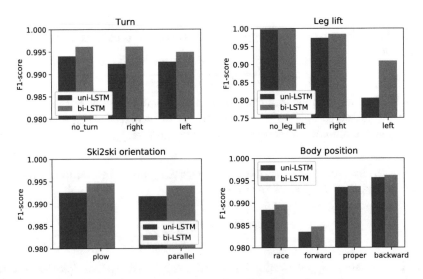

Fig. 5. F1-score of proposed models for each event

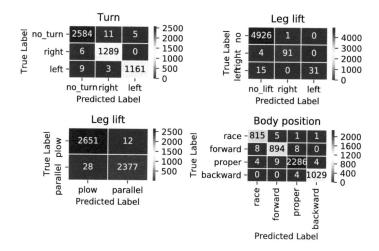

Fig. 6. Confusion matrices for unidirectional LSTM model for each event

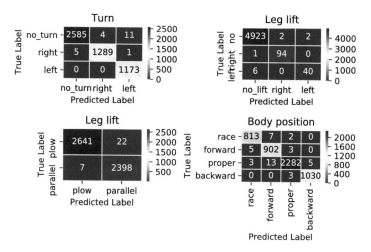

Fig. 7. Confusion matrices for bidirectional LSTM model for each event

6 Discussion

The proposed models identify quite well most of the activities, especially the most important one—turn. The imbalance of the data largely influenced the quality of the model responsible for recognizing leg lifts. This is due to the fact that there were few actions of that type. The solution to this problem is to gather more data and train the same model on a more complete set of data. Although the effectiveness of all models was decent, they were trained on data that were gathered on single ski slope. To further evaluate the reliability of the classifiers it would be reasonable to evaluate them on data with a different origin.

As expected, the use of future data improves the quality of the classification. For all types of activities, the model based on a bidirectional LSTM model gave better results. In the case of activities such as turning or determining the position of the body, the improvement was small—less than 1%. However, in the case of activity—leg lift, the improvement was significant from 92.55% to 96.41%, according to the F1 score.

7 Conclusions

This article presented and compare two LSTM architectures for skiing activity recognition. The bidirectional model was slightly better, however, before it will be used in the real system, it must be modified. Extending the base with next skiers or data from another slope should significantly increase the efficiency of the network performance. Another idea for modifications, which is planned in the future is to replace the LSTM based model with CNN based model. The first tests on this type of network gave interesting results, but not enough to include them in this work.

Acknowledgements. Publication supported by project *Innovative IT system to support alpine skiing and biathlon training, with the functions of acquisition of multimodal motion data, their visualization and advanced analysis using machine learning techniques, Snowcookie PRO*, Smart Growth Operational Program 2014–2020, POIR.01.01.01-00-0267/17 (2018–2020).

References

1. Aggarwal, J.K., Ryoo, M.S.: Human activity analysis: a review. ACM Comput. Surv. (CSUR) **43**(3), 16 (2011)
2. Ali, S., Basharat, A., Shah, M.: Chaotic invariants for human action recognition. In: 2007 IEEE 11th International Conference on Computer Vision, pp. 1–8. IEEE (2007)
3. Bachlin, M., Plotnik, M., Roggen, D., Maidan, I., Hausdorff, J.M., Giladi, N., Troster, G.: Wearable assistant for Parkinson's disease patients with the freezing of gait symptom. IEEE Trans. Inf Technol. Biomed. **14**(2), 436–446 (2010)
4. Chavarriaga, R., Sagha, H., Calatroni, A., Digumarti, S.T., Tröster, G., Millán, J.D.R., Roggen, D.: The opportunity challenge: a benchmark database for on-body sensor-based activity recognition. Pattern Recogn. Lett. **34**(15), 2033–2042 (2013)
5. Cho, H., Yoon, S.: Divide and conquer-based 1D CNN human activity recognition using test data sharpening. Sensors **18**(4), 1055 (2018)
6. Hnoohom, N., Mekruksavanich, S., Jitpattanakul, A.: Human activity recognition using triaxial acceleration data from smartphone and ensemble learning. In: 2017 13th International Conference on Signal-Image Technology & Internet-Based Systems (SITIS), pp. 408–412. IEEE (2017)
7. Jhuang, H.: A biologically inspired system for action recognition. Ph.D. thesis, Massachusetts Institute of Technology (2007)

8. Jiang, W., Yin, Z.: Human activity recognition using wearable sensors by deep convolutional neural networks. In: Proceedings of the 23rd ACM International Conference on Multimedia, pp. 1307–1310. ACM (2015)

9. Kingma, D.P., Ba, J.: Adam: a method for stochastic optimization (2014). arXiv preprint arXiv:1412.6980

10. Li, K., Zhao, X., Bian, J., Tan, M.: Sequential learning for multimodal 3D human activity recognition with long-short term memory. In: 2017 IEEE International Conference on Mechatronics and Automation (ICMA), pp. 1556–1561. IEEE (2017)

11. Murad, A., Pyun, J.Y.: Deep recurrent neural networks for human activity recognition. Sensors **17**(11), 2556 (2017)

12. Niebles, J.C., Fei-Fei, L.: A hierarchical model of shape and appearance for human action classification. In: 2007 IEEE Conference on Computer Vision and Pattern Recognition, pp. 1–8. Citeseer (2007)

13. Ordóñez, F., Roggen, D.: Deep convolutional and LSTM recurrent neural networks for multimodal wearable activity recognition. Sensors **16**(1), 115 (2016)

14. Snowcookie Team: System snowcookie® (2019). www.snowcookiesports.com

15. Turaga, P., Chellappa, R., Subrahmanian, V.S., Udrea, O.: Machine recognition of human activities: a survey. IEEE Trans. Circuits Syst. Video Technol. **18**(11), 1473 (2008)

16. Wang, J., Chen, Y., Hao, S., Peng, X., Hu, L.: Deep learning for sensor-based activity recognition: a survey. Pattern Recogn. Lett. **119**, 3–11 (2019)

17. Wang, L., Suter, D.: Recognizing human activities from silhouettes: motion subspace and factorial discriminative graphical model. In: 2007 IEEE Conference on Computer Vision and Pattern Recognition, pp. 1–8. IEEE (2007)

18. Zeng, M., Nguyen, L.T., Yu, B., Mengshoel, O.J., Zhu, J., Wu, P., Zhang, J.: Convolutional neural networks for human activity recognition using mobile sensors. In: 6th International Conference on Mobile Computing, Applications and Services, pp. 197–205. IEEE (2014)

19. Zhang, M., Sawchuk, A.A.: USC-HAD: a daily activity dataset for ubiquitous activity recognition using wearable sensors. In: Proceedings of the 2012 ACM Conference on Ubiquitous Computing, pp. 1036–1043. ACM (2012)

Recognition of Tennis Shots Using Convolutional Neural Networks Based on Three-Dimensional Data

Maria Skublewska-Paszkowska, Edyta Lukasik[✉], Bartłomiej Szydlowski, Jakub Smolka, and Pawel Powroznik

Lublin University of Technology, Nadbystrzycka 36D, 20-618 Lublin, Poland
{maria.paszkowska,e.lukasik,jakub.smolka,p.powroznik}@pollub.pl,
barszy94@gmail.com

Abstract. The article explores the possibilities of using convolutional neural networks to recognize the type of tennis shots. The study compares two architectures of such networks: Inception-v3 and MobileNet. Two datasets consisting of images of tennis players making an impact are analyzed in order to identify the type and phase of impact. The images in each of the collections were assigned to the respective five following classes: backhand preparation phase, backhand shot, forehand preparation phase, forehand shot and non-shot. In each of them there is a tennis player silhouette and a racket skeleton consisting of markers. Images of tennis players are generated from motion recordings made using motion capture technology. The networks were trained with different values of the learning rate. In addition, the network training time and match results for the best-trained models are presented. In view of the parameters considered, the MobileNet network has proved to be a better model.

Keywords: Convolutional Neural Networks · Inception-v3 · MobileNet · Tennis shots recognition

1   Introduction

Neural networks are currently one of the most dynamically developing branches of artificial intelligence. They are very widely used in such areas of life as: economy, medicine or IT. Among the most common applications of neural networks are the problems of image recognition and object recognition. Some of the systems based on them help to diagnose cancer by selecting images where the probability of lesions is high. Also in rehabilitation neural networks are a useful tool that allows for checking its effects and introduce any corrections to the treatment to improve results. In economics, neural networks are used to forecast sales, production or prices. Recently, there has been a significant development of machine learning, resulting in the creation of many specialized models of neural networks

© Springer Nature Switzerland AG 2020
A. Gruca et al. (Eds.): ICMMI 2019, AISC 1061, pp. 146–155, 2020.
https://doi.org/10.1007/978-3-030-31964-9_14

for solving specific problems. One of them involves the convolutional neural networks (CNN) used mainly for speech recognition and images. Convolutional nets are very good for recognizing individual objects and they can be replicated or scanned over large input images extremely cheaply and act as object spotters [8]. The aim of this article is to examine the possibility of using CNN to classify the type and phase of tennis player movement and to choose the best one. The analyzed cases of impact are: backhand preparation phase, backhand impact, forehand preparation phase, forehand shot and no shot. Two CNN architectures used for this study are: Inception-v3 [11] and MobileNet [6]. A network with better results is indicated.

2   Literature Review

Machine learning issues are becoming increasingly popular. This is evidenced by the increasing number of publications dealing with this topic. Currently, neural networks can be used for various purposes. CNNs are also tried as a tool for recognizing motion in video recordings. In [14] a method was proposed, in which at the beginning deep motion maps are generated and transferred to the model of the convolutional network. Images of various sports disciplines using the most popular databases were analyzed using convolutional neural networks. The UCF-Sport database [10] contains ten different disciplines, however tennis is not included. Base Sports-1M [7] also consists of videos which only specify the discipline. It is an extensive collection of one million videos assigned to 487 classes. 200,000 videos selected from it were processed using CNN and an increase in efficiency for the proposed architectures was obtained. There is a comprehensive THree dimEnsional TennIs Shots (THETIS) database containing 12 basic tennis shots captured by Kinect performed by 31 amateurs and 24 experienced players. The data are provided in 5 different synchronized forms (RGB, silhouettes, depth, 2D skeleton and 3D skeleton). The shots performed are: Backhand with two hands, Backhand, Backhand slice, Backhand volley, Forehand flat, Forehand slice, Forehand volley, Service flat, Service kick, Service slice and Smash [5]. The tennis movement became the subject of study in article [9]. The research was carried out on a set of RGB recordings from the THETIS database. The Inception model is used there, which was tasked with separating features. The separated features become data for learning the proposed network. In [15] the authors present an optical descriptor, which classifies the impact divided only into right and left turns with the use of SVM. In [4] the tennis shot classification is considered in three classes: shot, service and no strike. It uses recordings from the ACASVA Actions Dataset database [3], which are not publicly available. The algorithm that extracts features from each frame individually by using a well-known CNN named Inception [11] was proposed in [4]. The resulting sequences of features are then fed to a deep neural network consisting of three stacked long- and short-term memory units (LSTMs), a particular type of recurrent neural network (RNN). Actions were described using HOG3D features and for a transfer method based on feature re-weighting and a novel method based on feature translation and scaling was used. In none of the works on the tennis player's movement the impact phase is taken into account. In this article, the images were created using motion capture technology and divided into five categories, taking into account the type of motion and its phase.

3   The Concept of the Convolutional Network

In recent years, the recognition of images has gained in strength thanks to the use of CNN. Starting with LeNet-5 [8] CNNs have a standard structure, which is the convolution layers occurring after one or more fully-connected layers. For the majority of datasets, the number of layers and their sizes increased. The structure of the CNN has the structure of an acyclic graph [12], in which successive layers have an increasingly smaller number of nodes. The inputs of a given layer are connected to the previous layer. The neurons forming the CNN have the same values of weights and parameters, which makes them identical. Thanks to this, the network operates on fewer parameters making it more efficient. Shared weights enable the learning of features regardless of the location of the nodes in a given model and the type of input data. Connectivity between nodes in a convolution network is limited and this results in local connection patterns. The architecture of the CNN consists of three main layers: convolutional layer, pooling layer and full layer [2]. A convolution operation is performed in the convolution layer. It involves the processing of two functions resulting in a third function being a modification of one of the two previous functions. In the context of deep learning, this operation consists in applying an appropriate filter to the input data and computing a dot product. In this layer there are filters that can be described using four hyperparameters: the size of the receptive field, depth, stride and zero padding. The size of the receptive field determines the size of the filter. The depth determines the number of channels needed to describe a given image (in the case of an RGB image, the required number of channels is 3). The stride hyperparameter specifies the distance between neurons. The last parameter, zero-complementation, assigns the last (extreme) values of the zero-reception field. Each filter recognizes a feature. The recognized elements are more complex with each subsequent layer. In the pooling layer, a process occurs that is performed on feature maps that consists in aggregating multiple values into one. The new value can be determined on the basis of the average value (average-pooling), the maximum value (max-pooling), the minimum value (min-pooling) or the sum (sum-pooling). The use of pooling layer reduces the number of features. In addition, it ensures the invariance of translation – the model is able to generalize the feature more generously and does not attach so much importance to its location. This is one of the factors that prevents the network from overfitting [1]. The inner layers are partially connected, while the last layer called the full layer (most often being a perceptron) is a fully connected layer and is responsible for the final classification of a given image to the appropriate class.

3.1 Inception Architecture

So far, there are 4 versions of the Inception architecture from v1 to v4. Adding additional layers resulted in a significant increase in the network depth and drastically increased the computational costs of the convolution. The effective solution was to divide the large size filters at the same level, making the model grow wider, not deeper. Inception-v2 has 42 deep layers. The next version of Inception the v3 has the structure of the Inception-v2 model with several corrections, i.e. use was made of the RMSProp optimizer, label smoothing, over-adjustment and overfitting, factorization of the 7×7 filter and batch normalization, which reduces the internal change in covariance in auxiliary classifiers [13].

3.2 MobileNet Architecture

The architecture of the MobileNet network is based on depthwise separable convolutions. In traditional convergent layers, the kernel is applied to all of channels of the input image. Performing a convolution results in combining values from all input channels and creating an output image with one channel per pixel. In MobileNet network, this operation is similar to a standard convolution operation, because both filter data and create new feature maps; however, it is performed in two steps. The first is the deep convolution, and the second is point convolution. The deep convolution operation is performed on each channel separately. If the input image has three channels, the deep convolution results in an output image with three channels with its own weight sets. Then, point convolution is carried out, which only differs from the traditional one in the size of the kernel taking the dimensions of 1×1. This operation leads to connecting all output channels to create a new feature map. This approach results in faster network operation. In the case of a 3×3 filter, the proposed solution is about 8–9 times faster than the traditional one and a deeply separable convolution requires less computing power [6]. The model that is used for the research has 28 layers, the first being a standard 3×3 convolution layer and 13 layers of deep convolution. In addition, batch normalization and the ReLU6 activation function are applied. The last layer is so-called classification layer.

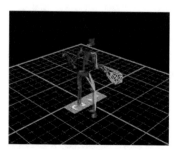

Fig. 1. Image with a tennis player's silhouette—forehand preparation.

4   Tests

The tests were carried out on a set of tennis players taking shots. The images were created from a recording of the tennis player's motion using a motion capture system. In each image there is a tennis player silhouette and a racket skeleton consisting of markers. A sample image is shown in Fig. 1. The study involved the use transfer learning and ready-made architectures of CNN. All layers in both networks had been overfitted with initial values obtained during the training in the ImageNet network, whereby architectures can learn new features faster. The following architectures were used for the research: Inception-v3 and MobileNet. In the Inception-v3 network, the last two layers were replaced with a custom classifier, while the top 6 layers were removed in the MobileNet model and replaced with a custom classifier. The aim of the study was to train the CNN so that it correctly classifies the types and phases of tennis player strokes. Two teaching sets of data were created to accomplish this task. The first one, which included images of three right-handed tennis players and the second one, apart from the three right-handed ones mentioned above, contained also one left-handed tennis player. The images in each of the collections were assigned to the respective five following classes: backhand preparation phase, backhand shot, forehand preparation phase, forehand shot and non-shot. The images were separated into five separate directories, where the name of the given folder marked the label of the class to which the image belonged. Then each of the sets was divided into training data, validation data and test data in a ratio of 60%/20%/20%. The first of the collections contained a total of 1692 photos and was divided into three subsets of data of the size of 1011/340/341. The second set had 2151 images and was divided into three data subsets sized 1293/428/430. The tests were carried out on a machine containing 2 GB of RAM and a 2 GB NVIDIA GeForce GTX 660M video card. Due to the relatively low computing power of the test computer, the size of training batches was 10, the size of the validation batches was 5 and the size of the test batches – 4. The used networks utilize ADAM optimizer (stochastic optimization method) and RMSprop with different values of the learning rate. The task of the optimization algorithm is to update the weights in each epoch on the basis of training data, while the learning rate controls the extent to which the network weights are adjusted taking into account gradient losses. In addition, cross-entropy was used as a function of loss, and the softmax activation function was applied in the highest prediction layer. Training steps have been selected so that the network in every epoch could use all available images. Network training was set to 20 epochs.

5   Results

In total, 24 training scenarios were conducted. Both networks were tested for a set of right-handed players and for a set with all players. Three values of the learning rate: 0.001, 0.0001 and 0.00001 were used. The best results were achieved for both MobileNet and Inception-v3 networks trained with the learning rate of

0.00001. The total time needed to train the network for 24 cases was 17 h 33 min. The accuracy and error functions, taking into account the type of learning sets (right-handed or all players) and the type of optimizer (Adam, RMSprop) for the MobileNet network are shown in Figs. 2, 3, 4 and 5, and for the Inception-v3 network in Figs. 6, 7, 8 and 9. Table 1 shows the training times and the effectiveness of both trained networks, including the dataset and the optimizer used for the learning rate of 0.00001 and 20 epochs.

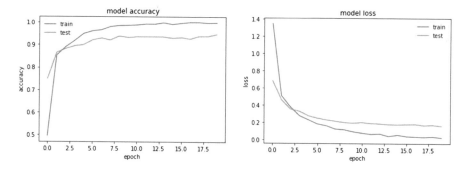

Fig. 2. Result of the MobileNet network training with the Adam optimizer and the learning rate of 0.00001 for right-handed players after 20 epochs.

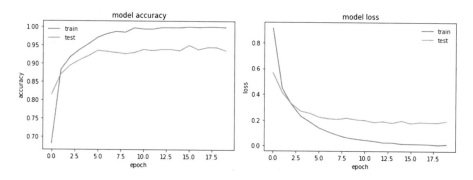

Fig. 3. Result of training the MobileNet network with the RMSprop optimizer and the learning rate of 0.00001 for right-handed players after 20 epochs.

6   Analysis of Results

For the presented training results, the smallest prediction error was obtained by the MobileNet network with the Adam optimizer. Regardless of the selected set of data (Figs. 2 and 4), the model along with subsequent epochs obtained ever smaller values of error. This proves that the problem is being generalized by the network. In the case of the same network and application of the RMSprop optimization function (Figs. 3 and 5), the error values stabilized after approximately

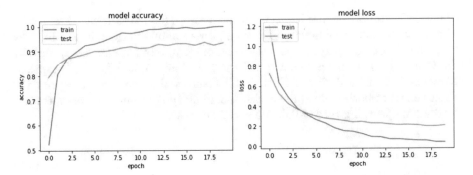

Fig. 4. Result of training the MobileNet network with the Adam optimizer and the learning rate of 0.00001 for all players after 20 epochs.

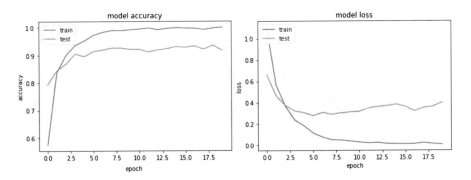

Fig. 5. Result of training the MobileNet network with the RMSprop optimizer and the learning rate of 0.00001 for all players after 20 epochs.

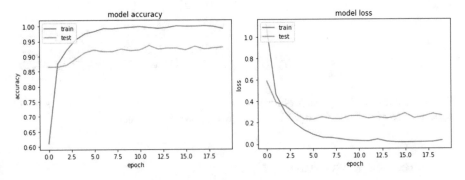

Fig. 6. Result of training the Inception-v3 network with the Adam optimizer and the learning rate of 0.00001 for right-handed players after 20 epochs.

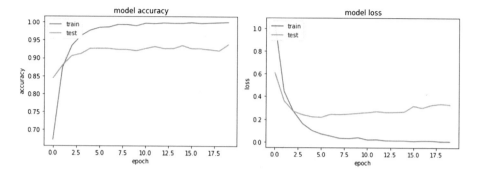

Fig. 7. Results of training the Inception-v3 network with the RMSprop optimizer and the learning rate of 0.00001 for right-handed players after 20 epochs.

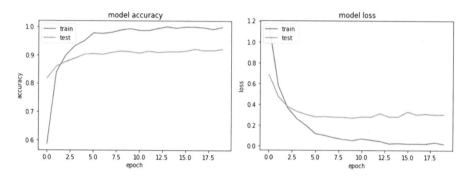

Fig. 8. Result of training the Inception-v3 network with the Adam optimizer and the learning rate of 0.00001 for all players after 20 epochs.

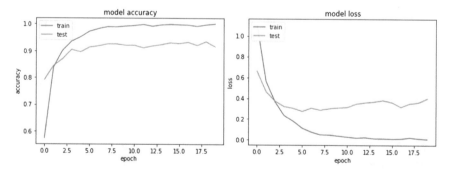

Fig. 9. Result of training the Inception-v3 network with the RMSprop optimizer and the learning rate of 0.00001 for all players after 20 epochs.

Table 1. Network training times for the learning rate of 0.00001 for 20 training epochs and the accuracy of the classification of test data.

Architecture name	Dataset	Optimiser	Training time [min]	Accuracy [%]
Inception-v3	Right-handed	Adam	60.26	94.7
		RMSprop	67.53	95.88
	Everyone	Adam	77.54	92.57
		RMSprop	75.66	94.65
MobileNet	Right-handed	Adam	25.41	95.6
		RMSprop	25.04	95.89
	Everyone	Adam	32.83	95.11
		RMSprop	32.28	95.81

5 epochs with a tendency to slightly increase. Also for the Inception-v3 network with the Adam optimization function (Figs. 6 and 8) the error value was the smallest, for both sets of data. However, in the case of Inception-v3, due to the complex network architecture and the use of the Adam optimizer, the error value decreases and stagnation occurs. Using the RMSprop optimizer (Figs. 7 and 9) initially reduces the error, but then the upward trend and possible problems with the exact generalization can clearly be seen. Despite the tendency of the Inception-v3 model to overfit in subsequent epochs, it was able to classify images from test data with over 94% accuracy. This is due to the use of the softmax activation function in the last prediction layer of the model. The accuracy of the model is evaluated on the basis of the cross entropy of the highest signal coming from the prediction layer and the regularity of image classification. The Inception-v3 network has a complex structure and contains a large number of layers, while the second MobileNet network tested has fewer layers, which is clearly reflected in the training times (Table 1). The MobileNet training time for the "right-handed" dataset is 2.5 times shorter compared to Inception-v3, and for the larger "everyone" collection, the difference was 2.35 times.

7 Summary

The article presents the results of research conducted on the detection of various types and phases of tennis player shots by means of deep machine learning. The results obtained indicate that CNN can be used to classify images containing tennis strokes. The obtained results may be helpful in the analyzing tennis players' technique. Two architectures were examined: Inception-v3 and MobileNet using one of two optimizers: Adam and RMSprop. For the proposed data set, the better choice is the architecture of the MobileNet network due to the time of its training and the accuracy obtained.

References

1. Boureau, Y.L., Ponce, J., LeCun, Y.: A theoretical analysis of feature pooling in visual recognition. In: Proceedings of the 27th International Conference on Machine Learning (ICML-10), pp. 111–118 (2010)
2. Ciresan, D.C., Meier, U., Masci, J., Gambardella, L.M., Schmidhuber, J.: Flexible, high performance convolutional neural networks for image classification. In: Twenty-Second International Joint Conference on Artificial Intelligence (2011)
3. De Campos, T., Barnard, M., Mikolajczyk, K., Kittler, J., Yan, F., Christmas, W., Windridge, D.: An evaluation of bags-of-words and spatio-temporal shapes for action recognition. In: 2011 IEEE Workshop on Applications of Computer Vision (WACV), pp. 344–351. IEEE (2011)
4. FarajiDavar, N., De Campos, T., Kittler, J., Yan, F.: Transductive transfer learning for action recognition in tennis games. In: 2011 IEEE International Conference on Computer Vision Workshops (ICCV Workshops), pp. 1548–1553. IEEE (2011)
5. Gourgari, S., Goudelis, G., Karpouzis, K., Kollias, S.: THETIS: three dimensional tennis shots a human action dataset. In: Proceedings of the IEEE Conference on Computer Vision and Pattern Recognition Workshops, pp. 676–681 (2013)
6. Howard, A.G., Zhu, M., Chen, B., Kalenichenko, D., Wang, W., Weyand, T., Andreetto, M., Adam, H.: MobileNets: efficient convolutional neural networks for mobile vision applications. arXiv preprint arXiv:1704.04861 (2017)
7. Karpathy, A., Toderici, G., Shetty, S., Leung, T., Sukthankar, R., Fei-Fei, L.: Large-scale video classification with convolutional neural networks. In: Proceedings of the IEEE Conference on Computer Vision and Pattern Recognition, pp. 1725–1732 (2014)
8. LeCun, Y., Boser, B., Denker, J.S., Henderson, D., Howard, R.E., Hubbard, W., Jackel, L.D.: Backpropagation applied to handwritten zip code recognition. Neural Comput. 1(4), 541–551 (1989)
9. Mora, S.V., Knottenbelt, W.J.: Deep learning for domain-specific action recognition in tennis. In: 2017 IEEE Conference on Computer Vision and Pattern Recognition Workshops (CVPRW), pp. 170–178. IEEE (2017)
10. Soomro, K., Zamir, A.R.: Action recognition in realistic sports videos. In: Computer Vision in Sports, pp. 181–208. Springer (2014)
11. Szegedy, C., Liu, W., Jia, Y., Sermanet, P., Reed, S., Anguelov, D., Erhan, D., Vanhoucke, V., Rabinovich, A.: Going deeper with convolutions. In: Proceedings of the IEEE Conference on Computer Vision and Pattern Recognition, pp. 1–9 (2015)
12. Szegedy, C., Vanhoucke, V., Ioffe, S., Shlens, J., Wojna, Z.: Rethinking the inception architecture for computer vision. In: Proceedings of the IEEE Conference on Computer Vision and Pattern Recognition, pp. 2818–2826 (2016)
13. Toward Data Science: A simple guide to the versions of the inception network. https://towardsdatascience.com/a-simple-guide-to-the-versions-of-the-inception-network-7fc52b863202. Accessed 1 Mar 2019
14. Wang, P., Li, W., Gao, Z., Zhang, J., Tang, C., Ogunbona, P.: Deep convolutional neural networks for action recognition using depth map sequences. arXiv preprint arXiv:1501.04686 (2015)
15. Zhu, G., Xu, C., Gao, W., Huang, Q.: Action recognition in broadcast tennis video using optical flow and support vector machine. In: European Conference on Computer Vision, pp. 89–98. Springer (2006)

On Unsupervised and Supervised Discretisation in Mining Stylometric Features

Urszula Stańczyk[✉]

Silesian University of Technology, Akademicka 2A, 44-100 Gliwice, Poland
urszula.stanczyk@polsl.pl

Abstract. Writing styles can be described by stylometric features. They are quantitative in nature, often continuous, which either limits techniques used in mining to those capable of calculations on this form, or adds discretisation to initial pre-processing of data. The paper describes research on unsupervised and supervised discretisation applied in the stylometric domain for the task of authorship attribution. The recognition of authorship is executed as classification performed by chosen inducers capable of operating on both continuous and categorical attributes.

Keywords: Stylometry · Feature · Authorship attribution · Unsupervised discretisation · Supervised discretisation

1 Introduction

Writing styles can be defined and described by characteristic features reflecting both conscious and sub-conscious language preferences and traits of authors [1]. These elements need to be detectable regardless of a subject content of any given text, difficult to imitate with premeditation, and should be relatively easy to operate on in some automatic manner. Textual analysis often refers to quantitative descriptors of lexical or syntactic type [3], for example giving frequencies of occurrence for popular function words, used punctuation marks.

Numerical characteristics of stylometric features enable execution of authorship attribution as a task of classification with the help of some statistic-based calculations or machine learning approaches [9]. Data mining techniques to be used either need to be able to work in continuous input space, or require introduction of an additional step in pre-processing of data, dedicated to discretisation.

Discretisation transforms real-valued variables into nominal ones, by performing some reduction of data [5]. Methods are divided into several categories, one of which makes a distinction of class information. When such information has bearing upon the construction of intervals and selected cut-points, a method is named *supervised*. If it is disregarded, then this method is called *unsupervised*.

In this paper there are presented results from research concerning two discretisation methods: supervised Kononenko [10], and unsupervised equal frequency binning, applied, separately and in combination, in the stylometric task

© Springer Nature Switzerland AG 2020
A. Gruca et al. (Eds.): ICMMI 2019, AISC 1061, pp. 156–166, 2020.
https://doi.org/10.1007/978-3-030-31964-9_15

of authorship attribution. Effects of discretisation were studied in the context of performance of selected inducers, evaluated with test sets.

The content of the paper is organised as follows. Section 2 describes briefly characteristics of stylometric features. Section 3 contains comments on discretisation approaches. Section 4 provides explanation for the setup of experiments, and their results are discussed in Sect. 5. Section 6 concludes the paper.

2 Characteristics of Stylometric Features

When a writing style is described by characteristic features [4], the problem of attributing authorship with sufficient level of reliability can be treated as classification. Such reasoning leads to application of data mining methods [11, 15].

To obtain reliable textual descriptors, construction of suitable text samples is required. The samples need to be of sufficient length, which should also be comparable [6]. In the standard approach larger works are divided into smaller parts of defined size. Such procedure allows to increase the cardinalities of sets and observe style variations, which otherwise could be lost in more general characteristics. As writing styles of authors of the same gender show some similar traits [16], in order to obtain unbiased conclusions it is a good practice to separate writers of the opposite sex, and construct disjoint sets of samples.

Since text samples can be grouped basing on larger documents they are part of, samples included in one group show closer similarity than those from different groups. As a result, cross-validation, popularly employed for performance evaluation of classification systems, tends to return falsely high accuracy of predictions [2]. Therefore, employing test sets, including samples based only on separate works, is more reliable. It is also best to ensure balance of classes [17].

Text samples constitute a base for which next linguistic observations are formulated, often referring to some quantitative characteristics [13], such as averages, distributions, frequencies. These elements return continuous-valued features and from all available some arbitrary subset can be chosen, or other methodologies used in the process of feature selection, for example ranking [19].

3 Unsupervised and Supervised Discretisation

Discretisation process is dedicated to reduction of data. It translates input continuous domain of attributes into categorical [7,12]. When, in calculations of cut-points between constructed intervals, information on considered classes is taken into account, the procedures are called supervised. When only values of attributes are studied, regardless on classes, the process is unsupervised.

Supervised discretisation is often preferred over unsupervised. Kononenko's algorithm [10] begins its work by assigning a single interval to represent the whole range of values for each processed variable. Next these bins are analysed to verify whether they should be split into sub-ranges. Kononenko showed that the candidates for possible cut-points can be accepted when the inequality (1), based on Minimum Description Length (MDL) principle, holds true. This stopping criterion is given by:

$$\log N_T + \sum_j \log \begin{pmatrix} N_{a_j} \\ N_{d_1 a_j} \dots N_{d_k a_j} \end{pmatrix} + \sum_j \begin{pmatrix} N_{a_j} + k - 1 \\ k - 1 \end{pmatrix}$$

$$< \log \begin{pmatrix} N \\ N_{d_1} \dots N_{d_k} \end{pmatrix} + \log \begin{pmatrix} N + k - 1 \\ k - 1 \end{pmatrix}, \tag{1}$$

where N is the number of instances in the training set, N_T gives the number of considered cut-points T, and N_{d_i} specifies the number of examples with the decision d_i. N_{a_x} denotes the number of instances for the given attribute a with the value x, and $N_{d_i a_y}$ the number of examples in the learning set with the decision d_i, for which the value of the given attribute a is y.

The search for cut-points is performed recursively as long as the inequality (1) is satisfied. If the inequality is false for all candidate cut-points from the very beginning of the partitioning procedure, as a result for the considered attribute the initial single interval remains the sole representative for all of its values.

Kononenko approach works in the context of a given input set, which can be problematic while dealing with several disjoint input sets of examples. In this case it is quite likely that not only different categories are constructed, but their numbers can also significantly vary, when for some variable in one set a single interval is found and in the other set several bins are defined [18].

Unsupervised discretisation allows to form requested numbers of intervals for considered attributes even for disjoint sets of instances. Equal frequency binning belongs with this category of methods. Alike Kononenko algorithm, it does not cause the uniform partitioning of the input continuous space, instead the bins reflect grouping of existing data points. The number of constructed intervals is the input parameter, and such split points are chosen which ensure that the numbers of attribute values included are the same for all bins.

4   Conditions of Performed Experiments

The experiments began with preparation of input data sets with continuous attributes. For these sets and chosen inducers the reference performance for comparisons was established. Then all sets were discretised and resulting classification accuracy verified. Test conditions are detailed in the following sections.

4.1   Input Stylometric Data Sets

For recognition two pairs of writers were selected, E. Wharton and M. Johnston for female, and J.F. Curwood and J. London for male writer data set. Their works were divided into 3 separate groups, as a base for samples to be included in the training, and 2 test sets. The novels were further divided into smaller parts, over which frequencies of occurrence for chosen markers were calculated.

The employed textual descriptors were of syntactic and lexical nature, obtained by referring to punctuation marks and function words often used in English language. Initially, the set of 100 features was formed, next reduced with the help of ranking procedures (by taking the highest ranking attributes), which resulted in the final set of 24 stylometric characteristic features.

4.2   Classification Systems Used

As the research focused on the influence of discretisation methods on classification accuracy in the process of pattern recognition, for evaluation of results three popular inducers were selected, namely: k-Nearest Neighbour (kNN), Naive Bayes (NB), and Random Forest (RF), implemented in WEKA [20] workbench.

Naive Bayes is a probabilistic classifier that relies on Bayes rule of conditional entropy, which shows similarity to calculations executed for supervised discretisation process. k-Nearest Neighbour algorithm makes the decision with respect to the class for a given sample basing on the majority of votes from its closest neighbours. Random Forest builds decision trees for randomly selected subsets of training samples, and selects the class through aggregation of votes, combining aggregation with randomisation [8,14].

All three classification systems can operate on both numerical and categorical attributes, and all were used with default settings. Their performance for both data sets, while still operating in the continuous domain, is listed in Table 1. The column AvgT denotes the average calculated over two test sets for an inducer, while AvgD gives the average over all classification systems for a data set.

Table 1. Performance of selected inducers for attributes in continuous domain [%].

Dataset	Classifiers and test sets									AvgD
	kNN			NB			RF			
	T1	T2	AvgT	T1	T2	AvgT	T1	T2	AvgT	
Female writer	96.67	94.44	95.56	95.56	91.11	93.33	97.78	94.44	96.11	95.00
Male writer	93.33	92.22	92.78	88.89	93.33	91.11	90.00	93.33	91.67	91.85

4.3   Employed Discretisation Approaches

In the research three different ways of discretisation, corresponding to three parts of performed experiments, were used: supervised Kononenko, unsupervised equal frequency binning, and a combination of these two approaches.

In the first part of experiments Kononenko method was employed to process independently all input sets. Due to this independent transformation and specifics of the method, the input continuous space was translated into categorical, closely relating the distribution of existing data points for each considered set separately. As a consequence, for the same variables in different sets varying numbers of intervals were constructed, including cases of single bins for some attributes. The maximal number of intervals created was 4.

Unsupervised equal frequency binning was applied in the second part of experiments. In this approach the number of bins to be constructed is the input parameter. This number varied from 2 to 100 with a step of 1. The number of instances included in a bin depends on the number of available unique values for a variable, thus sets with the same input parameters but different cardinalities receive the same numbers of bins but definitions of cut-points differ, which obviously applies to the distinction of learning and test sets, the samples of the former most often outnumbering the latter.

For the third part of experiments there were taken sets obtained previously through supervised discretisation, but they were further processed. All attributes that in discretisation of learning sets received only a single bin for representation were transformed in all considered sets (both learning and test), by applying to their continuous domains equal frequency binning with the range of bins constructed from 2 to 10. Each version of a learning set was then tested with each version of a test set, as commented in the following section.

5   Discussion of Obtained Results

The results from the first stage of experiments, dedicated to application of supervised Kononenko discretisation, are listed in Table 2. When the performance averaged over all test sets and inducers is compared with the previously obtained for continuous valued features, it turns out that for female writer data set it became degraded, while for male writer data set it was improved.

Table 2. Performance of inducers for attributes discretised by Kononenko method [%].

Dataset	Classifiers and test sets									AvgD
	kNN			NB			RF			
	T1	T2	AvgT	T1	T2	AvgT	T1	T2	AvgT	
Female writer	94.44	76.67	85.56	96.67	85.56	91.12	93.33	92.22	92.78	89.82
Male writer	87.78	93.33	90.56	96.67	95.56	96.12	93.33	93.33	93.33	93.37

All classifiers returned worsened results in case of female writer recognition, for male only kNN, and NB and RF performed better than before. In fact the lowest classification accuracy for both data sets was reported by kNN. These results were caused (to some extent) by construction of varying numbers of intervals for the same attributes through independent discretisation of data sets.

In the second phase of tests unsupervised equal frequency binning was applied to all features. The number of required bins varied from 2 to 100. Table 3 lists obtained minimum, maximum, and calculated overall average for each inducer separately, and also for each data set (that is over all classifiers). As for supervised discretisation, equal frequency binning caused on average worsened results for female writer recognition, and enhanced for male data set, however, both averages were higher than those reported for Kononenko method. It was due to inflexible input parameter for all processed sets building the same numbers of bins. This made domains of attributes closer than could be obtained by independent supervised discretisation.

Table 3. Performance of inducers for attributes discretised by equal frequency binning method [%].

Dataset	Classifiers and test sets									AvgD
	kNN			NB			RF			
	Min	Max	Avg	Min	Max	Avg	Min	Max	Avg	
Female writer	88.33	93.89	91.94	88.89	94.44	90.40	92.22	99.44	97.38	93.24
Male writer	90.56	95.56	93.72	93.33	96.67	94.74	88.33	97.78	94.33	94.26

The trends in performance are shown in Fig. 1. For each inducer there are displayed averages calculated for the two test sets in relation to the number of bins considered. For female writer data set these trends for the considered classifiers are rather clearly separated, with the best shown by RF (at maximum very close to perfect) and the worst results due to NB. For male writers the results from all inducers were relatively close to each other, with the lowest minimum due to RF, which, however, was also responsible for the highest maximum.

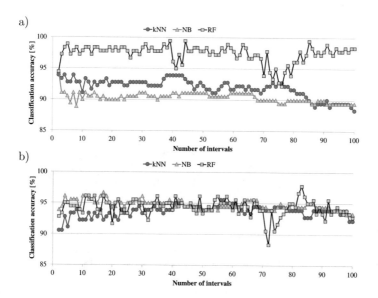

Fig. 1. Averaged classification accuracy [%], for attributes discretised by equal frequency binning for: (a) female writer dataset, (b) male writer dataset.

The results from the third step of experiments, focused on combined supervised with unsupervised discretisation applied to features, are shown separately for each data set and each inducer, with averaging over test sets, respectively in Fig. 2 for female writer data set, and in Fig. 3 for male writer data set.

As previously described, variables which in supervised discretisation of training sets received single intervals, were further processed by equal frequency binning applied to their continuous domains in all sets. The number of bins ranged from 2 to 10, which resulted in 9 more versions of each set, all versions tested with each rather. In the plots axis X displays the number of bins for additionally transformed attributes for the learning sets, while series refer to the number of intervals in test sets. Series 1 correspond to sets obtained by pure supervised Kononenko discretisation, without additional unsupervised processing.

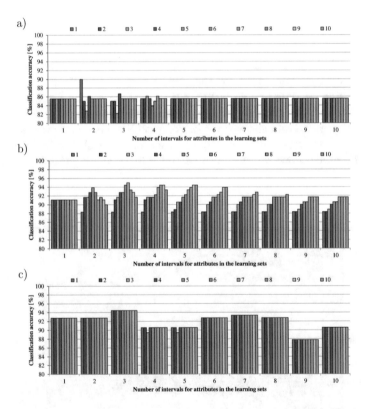

Fig. 2. Averaged classification accuracy [%] for attributes discretised by combined supervised Kononenko and unsupervised equal frequency binning for female writer dataset, for: (a) k-Nearest Neighbours, (b) Naive Bayes, (c) Random Forest.

For female writer recognition only NB indicated noticeable dependence on a particular combination of numbers of bins requested for learning and test sets, and it also reported cases of improvement in classification. In can be observed that enhanced results were obtained when more intervals were formed in test sets than in learning sets for the attributes selected for extra transformations. For the other two inducers there were fewer conditions for enhanced prediction and almost no dependence on additional processing of test sets.

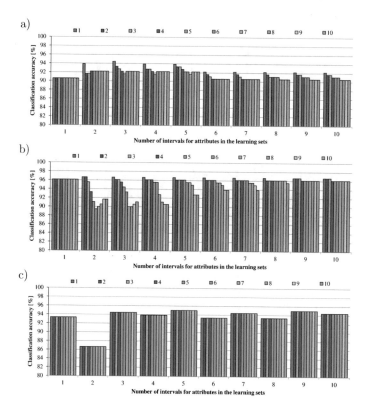

Fig. 3. Averaged classification accuracy [%] for attributes discretised by combined supervised Kononenko and unsupervised equal frequency binning for male writer dataset, for: (a) k-Nearest Neighbours, (b) Naive Bayes, (c) Random Forest.

For male writers for all inducers some further improvement after combined discretisation was detected. For kNN and NB the best results were obtained with processed learning sets, while limiting test sets to supervised discretisation. For RF the performance depended only on additional processing of training samples and showed no changes with respect to transformations for test sets.

The charts presented in Fig. 4 allow to analyse the effect of combined discretisation with respect to each data set, as it displays the performance averaged for all three inducers. For both female and male writer data sets additional processing gave chance for improvement when compared with purely supervised discretisation, which confirms the merit of such methodology. However, for male writers for a version of a learning set increasing the number of bins in test sets brought worsening of recognition, while for female writers there was an opposite trend: with more intervals in test sets the reported results were better.

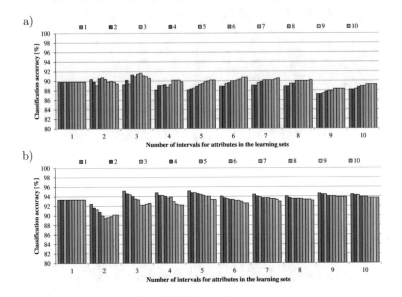

Fig. 4. Performance [%] averaged over all considered classifiers for attributes discretised by combined supervised Kononenko and unsupervised equal frequency binning for: (a) female writer dataset, (b) male writer dataset.

6   Conclusions

In the paper there are described results from research on discretisation approaches applied in mining of stylometric characteristic features. In two tasks of binary authorship attribution with balanced data, the performance of three popular inducers (k-Nearest Neighbour, Naive Bayes, Random Forest) was evaluated by separate test sets. These classifiers are capable of working both on numerical and nominal attributes, and their powers in these cases were compared.

There were investigated three discretisation procedures applied to data: supervised Kononenko, unsupervised equal frequency binning, and a combination of these two algorithms. Performed experiments indicate that when there are independently discretised several separate data sets including the same attributes, supervised discretisation can result in worse performance than unsupervised, which, however, can be improved to some extent by additional transformations of attributes for which only single intervals were defined.

Acknowledgements. The research presented was executed within the project BK/RAu2/2019 at the Silesian University of Technology, Gliwice, Poland.

References

1. Argamon, S., Burns, K., Dubnov, S. (eds.): The Structure of Style: Algorithmic Approaches to Understanding Manner and Meaning. Springer, Berlin (2010)
2. Baron, G.: Comparison of cross-validation and test sets approaches to evaluation of classifiers in authorship attribution domain. In: Czachórski, T., Gelenbe, E., Grochla, K., Lent, R. (eds.) Proceedings of the 31st International Symposium on Computer and Information Sciences. Communications in Computer and Information Science, vol. 659, pp. 81–89. Springer, Cracow (2016)
3. Burrows, J.: Textual analysis. In: Schreibman, S., Siemens, R., Unsworth, J. (eds.) A Companion to Digital Humanities. Blackwell, Oxford (2004)
4. Craig, H.: Stylistic analysis and authorship studies. In: Schreibman, S., Siemens, R., Unsworth, J. (eds.) A Companion to Digital Humanities. Blackwell, Oxford (2004)
5. Dougherty, J., Kohavi, R., Sahami, M.: Supervised and unsupervised discretization of continuous features. In: Machine Learning Proceedings 1995: Proceedings of the 12th International Conference on Machine Learning, pp. 194–202. Elsevier (1995)
6. Eder, M.: Does size matter? Authorship attribution, small samples, big problem. Digit. Sch. Hum. **30**, 167–182 (2015)
7. García, S., Luengo, J., Sáez, J.A., López, V., Herrera, F.: A survey of discretization techniques: taxonomy and empirical analysis in supervised learning. IEEE Trans. Knowl. Data Eng. **25**(4), 734–750 (2013)
8. Han, J., Kamber, M., Pei, J.: Data Mining: Concepts and Techniques. Morgan Kaufmann, Burlington (2011)
9. Jockers, M., Witten, D.: A comparative study of machine learning methods for authorship attribution. Lit. Linguist. Comput. **25**(2), 215–223 (2010)
10. Kononenko, I.: On biases in estimating multi-valued attributes. In: Proceedings of the 14th International Joint Conference on Artificial Intelligence, IJCAI 1995, vol. 2, pp. 1034–1040. Morgan Kaufmann Publishers Inc. (1995)
11. Koppel, M., Schler, J., Argamon, S.: Computational methods in authorship attribution. J. Am. Soc. Inform. Sci. Technol. **60**(1), 9–26 (2009)
12. Kotsiantis, S., Kanellopoulos, D.: Discretization techniques: a recent survey. GESTS Int. Trans. Comput. Sci. Eng. **32**(1), 47–58 (2006)
13. Peng, R., Hengartner, H.: Quantitative analysis of literary styles. Am. Stat. **56**(3), 15–38 (2002)
14. Quinlan, R.: C4.5: Programs for Machine Learning. Morgan Kaufmann Publishers, San Mateo (1993)
15. Stamatatos, E.: A survey of modern authorship attribution methods. J. Am. Soc. Inform. Sci. Technol. **60**(3), 538–556 (2009)
16. Stańczyk, U.: Recognition of author gender for literary texts. In: Czachórski, T., Kozielski, S., Stańczyk, U. (eds.) Man-Machine Interactions 2. AISC, vol. 103, pp. 229–238. Springer, Berlin (2011)
17. Stańczyk, U.: The class imbalance problem in construction of training datasets for authorship attribution. In: Gruca, A., Brachman, A., Kozielski, S., Czachórski, T. (eds.) Man-Machine Interactions 4. AISC, vol. 391, pp. 535–547. Springer, Berlin (2016)
18. Stańczyk, U.: Evaluating importance for numbers of bins in discretised learning and test sets. In: Czarnowski, I., Howlett, J.R., Jain, C.L. (eds.) Intelligent Decision Technologies 2017: Proceedings of the 9th KES International Conference on Intelligent Decision Technologies (KES-IDT 2017) – Part II, vol. 73, pp. 159–169. Springer (2018)

19. Stańczyk, U., Zielosko, B., Żabiński, K.: Application of greedy heuristics for feature characterisation and selection: a case study in stylometric domain. In: Nguyen, H., Ha, Q., Li, T., Przybyla-Kasperek, M. (eds.) Proceedings of the International Joint Conference on Rough Sets, IJCRS 2018. Lecture Notes in Computer Science, vol. 11103, pp. 350–362. Springer, Quy Nhon (2018)
20. Witten, I., Frank, E., Hall, M.: Data Mining: Practical Machine Learning Tools and Techniques, 3rd edn. Morgan Kaufmann, Burlington (2011)

Bio-Data and Bio-Signal Analysis

LCR-BLAST—A New Modification of BLAST to Search for Similar Low Complexity Regions in Protein Sequences

Patryk Jarnot[1]($\boxtimes$), Joanna Ziemska-Legięcka[2], Marcin Grynberg[3], and Aleksandra Gruca[1]

[1] Institute of Informatics, Silesian University of Technology, Gliwice, Poland
{patryk.jarnot,aleksandra.gruca}@polsl.pl
[2] Faculty of Mathematics, Informatics and Mechanics, University of Warsaw, Warsaw, Poland
m.ziemska@gmail.com
[3] Institute of Biochemistry and Biophysics, Polish Academy of Sciences, Warsaw, Poland
greenb@ibb.waw.pl

Abstract. Low Complexity Regions (LCRs) are fragments of protein sequences that are characterized by a small diversity in amino acid composition. LCRs could play important roles in protein functions or they could be relevant to protein structure. However, for many years, low complexity regions were ignored by the scientific community which resulted in lack of algorithms and tools that could be used to analyze this specific type of protein sequences.

Recently, researchers became interested in the so-called dark proteome and studies on such kind of proteins revealed that a vast amount of them include LCRs. Therefore, there is an urgent need to adapt existing methods or develop new ones that could be useful for analysis of LCRs, especially in the context of their functional roles in protein sequences.

In this paper, we present LCR-BLAST which is a new modification of BLAST designed to search for similarities among LCRs. This modification consists of the following elements: turning on short sequence parameters, turning off compositional based statistics, applying our own version of identity scoring matrix and replacing E-value with mean-score statistics. In order to evaluate the performance of our new modification, we compare the number of similar pairs found by LCR-BLAST with performance of a standard BLAST tool and of BLAST with a specific set of parameters for compositionally biased and short sequences. We show that our new method provides a robust and balanced solution for searching for similarities among LCRs.

Keywords: Low complexity regions · Protein sequences · Sequence similarity · BLAST

© Springer Nature Switzerland AG 2020
A. Gruca et al. (Eds.): ICMMI 2019, AISC 1061, pp. 169–180, 2020.
https://doi.org/10.1007/978-3-030-31964-9_16

1   Introduction

Low Complexity Regions (LCRs) are fragments of protein sequences that are characterized by a small diversity in amino acid composition. Analysis of protein sequence databases reveals that about 14% [9] of proteins contain LCRs and that these LCRs could play important roles in protein function or they could be relevant to protein structure [2]. However, for many years, low complexity regions were largely ignored by the scientific community as irrelevant parts of sequences. This resulted in lack of algorithms and tools that could be used to analyze this specific type of protein sequences.

Recently, researchers got interested in the so-called dark proteome, which can be defined by intrinsically disordered proteins or proteins that contain intrinsically disordered regions [3,8]. Studies on such kind of proteins revealed that a vast amount of them include low complexity regions in their sequences. Therefore, there is an urgent need to adapt existing methods or to develop new ones that could be useful for analysis of LCRs, especially in the context of their functional roles in protein sequences.

Searching for similarities among protein sequences have always been an important tool in biology that allowed researchers to predict and conclude about protein function from the sequence data alone. However, all statistical models that are used to find and compare similarities among protein sequences are designed to process and analyze high complexity regions (HCRs) of protein sequences. Therefore in this study, we focus on analyzing how BLAST (the Basic Local Alignment Search Tool) [1], which is the most popular and widely used tool for searching for similar protein sequences, can be applied for the task of searching for similarities among low complexity parts of the proteins. BLAST is designed to find regions of local similarity between two sequences by comparing their similarity over the database of sequences and calculating the statistical significance of matches. BLAST can be used to conclude about functional and evolutionary relationships between sequences.

In this study, we present LCR-BLAST which is a new modification of BLAST designed to search for similarities among LCRs. This modification consists of the following elements: turning on short sequence parameters, turning off compositional based statistics, applying our own version of identity scoring matrix and replacing e-value with mean-score statistics. In order to evaluate the performance of our new modification we compare the number of similar pairs found by LCR-BLAST with the performance of a standard BLAST tool and three versions of BLAST modified with specific sets of parameters designed for short, compositionally biased sequences. In the result section, we show that default BLAST masks a lot of hits, good candidates for LCRs while BLAST with short sequence and compositionally biased statistics parameter settings finds too many hits which could include a lot of mismatches. We conclude that our new method provides a balanced solution for searching for similarities among LCRs.

2   Methods

In this section, we present the BLAST method, its different parameter settings and workflow of our method. We discuss the possible influence on different parameters on the results of analyses. Modifications of BLAST parameters analyzed in this work are presented in Table 1.

Table 1. Modifications of BLAST parameters analyzed in this work

Name	Description
BLAST	Default parameters
SHORT-BLAST	Short sequences turned on
COMPOSITION-BLAST	Composition based scoring matrix adjustment turned off
SHORT-COMPOSITION-BLAST	Short sequences turned on
	and composition based scoring matrix adjustment turned off
LCR–BLAST	Short sequences turned on,
	composition based scoring matrix adjustment turned off
	identity scoring matrix and mean score statistics

BLAST is a popular tool for searching for evolutionarily related proteins. It uses the Smith-Waterman algorithm [11] to calculate the similarity score between two sequences. BLAST uses BLOSUM [6] or PAM [5] matrices to search for homologous proteins and in order to improve the results, BLAST has several techniques to mask LCR or to decrease their significance. User can mask LCRs using SEG tool [12] or use one of the compositionally biased score adjustments to recalculate a scoring matrix. Result hits are selected using e-value statistics that reflect how many times one can expect a hit with a similar score just by chance.

2.1   Default Parameters for Short Sequences

Regardless of the length of input sequence BLAST uses default parameters for long sequences to find evolutionarily related sequences. That is due to the fact that most of the sequences are rather long, whereas LCRs are mainly short.

Distribution of low complexity regions collected from UniProtKB/Swiss-Prot [4] (version form April, 2019) by SEG tool is shown in Fig. 1. Average sequence length is 24.03 with standard deviation 26.37. The shortest sequence SEG was able to detect has 8 amino acids whereas the longest has 2302 amino acids. First, second and third quartilea are 16, 20 and 26 respectively. By analyzing these results we notice that LCRs are mainly short sequences. Considering that we decided to use parameters for short sequences as a default setting for LCR-BLAST.

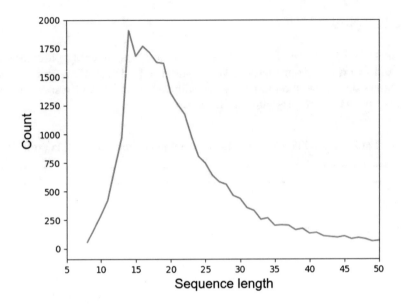

Fig. 1. Distribution of lengths of LCRs in UniProtKB/Swiss-Prot

2.2   Composition Based Statistics

The situation in which two non-homologous sequences contain strongly similar LCRs occurs quite often. Because BLAST uses local alignment to search for homologous sequences, it may find many sequences which are not homologous but contain similar LCRs. In order to overcome this limitation, composition based statistics were introduced to BLAST. This statistics uses the Eq. 1 to recalculate scoring matrix in order to decrease the score of frequently occurring amino acids in a query sequence.

$$P_i P_j exp(\lambda S_{ij}) = P_i^* P_j exp(\lambda S_{ij}^*) = Q_{ij}, \tag{1}$$

where P_i and P_j are frequencies of occurrence of the amino acid in the query sequence, λ is a normalizing parameter, S_{ij} is the score in matrix at a given position and S_{ij}^* is a modified score.

When scoring matrices are created, they contain real numbers, because in computing it is more convenient to use integer numbers. To convert these real numbers to integers, the values are multiplied by a factor to reduce the loss of information. For this type of normalization λ parameter is used.

2.3   Identity Scoring Matrix

BLAST developed two different families of scoring matrices which are BLOSUM and PAM. Both of them are designed to search for homologous sequences, however LCRs can be both homologous and analogous. For this reason, we have to reject these matrices. As a replacement, we propose a similarity scoring matrix which is defined as in Eq. 2.

$$S(a_i, b_j) = \begin{cases} +6, a_i = b_j \\ -6, a_i \neq b_j \end{cases} \tag{2}$$

The score is calculated using the Smith-Waterman algorithm. There are three variables that influence it: gap open, gap extend and the scoring matrix. These values are integer and have to be adjusted to each other. If the scoring matrix assigns one point for match and minus one for mismatch, there will be no possibility to set gap open and extend penalties lower than mismatch. That is why values of identity scoring matrix have to be more distant from zero. Therefore, by empirical analysis we propose 6 for match and −6 for mismatch to calculate similarity score of alignment.

2.4   Mean Score

The main statistics to select and sort BLAST results is e-value [7]. This statistics is described as *parameter that describes the number of hits one can get by chance when searching a database of a particular size*. It decreases exponentially as the Score (S) of the match increases. However, the number of occurrences of LCRs with similar amino acid patterns is relatively high. That is why we search for similar sequences regarding their amino acid composition similarity and we do not take into account a probability that the sequence can occur in a database. For that reason we decided to replace e-value by mean score statistics M which is denoted by the Eq. 3:

$$M = \frac{S}{L}, \tag{3}$$

where S denotes Score and L is the length of an alignment.

Mean score statistics is independent of sequence length. It can be found in the literature that LCRs can have a function if their length reaches a particular size [10]. In order to do that biologists have to collect similar LCRs with different lengths. Proposed statistic allows them to find similar sequences which vary in length. In case we want to be length dependent we can always use score S to refine the results.

2.5   Workflow

We used the UniProtKB/Swiss-Prot database to find similar subsequences of proteins. In the database we identified LCRs using SEG with strict parameters (Window size $W = 15$, $K1 = 1.5$, $K2 = 1.8$). Then we used BLAST and each modification of it to identify similar sequences. As a result, we obtained five lists of similar pairs of protein subsequences. The whole workflow is presented in Fig. 2.

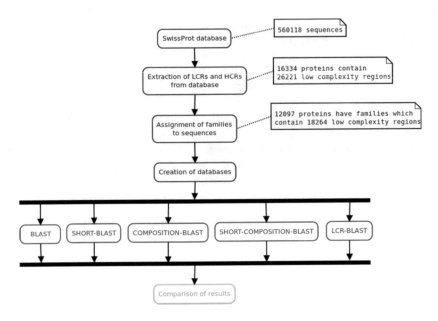

Fig. 2. The workflow of experimental analysis

3   Data Analysis and Results

In this section, we present a comparison of the results obtained with the use of different methods. The overlap among the results from all methods is presented in Figs. 3 and 4 presents how the results obtained with default BLAST overlap with results obtained with BLAST with different parameter settings and finally with LCR-BLAST.

The main factor that influences the difference between number of proteins with similar sequences is *composition based statistics*. When composition based adjustment of the scoring matrix is applied then it strongly decreases score value for frequently occurring amino acids. Therefore, BLAST results do not include homopolymers which are special and frequent cases of LCRs.

In Fig. 4d we present the Venn diagram for LCR-BLAST and BLAST. We can see that LCR-BLAST found much more similar LCRs than BLAST, that is 136753 and 4173 respectively. However, regular BLAST found some sequences which were not found by LCR-BLAST. If we analyze these sequences, we can notice that these sequences are out of our mean score threshold, which means that they are not similar according to what we assumed for LCR-BLAST.

Distribution of alignment lengths found by each method is presented in Fig. 5. It shows that BLAST, SHORT-BLAST, COMPOSITION-BLAST and SHORT-COMPOSITION-BLAST do not find sequences shorter than 9 amino acids. The main reason is the length dependence of e-value. Intuitively shorter sequences should be more likely low complexity regions just by chance. We can observe this situation in case of LCR-BLAST alignments. In case of other methods the

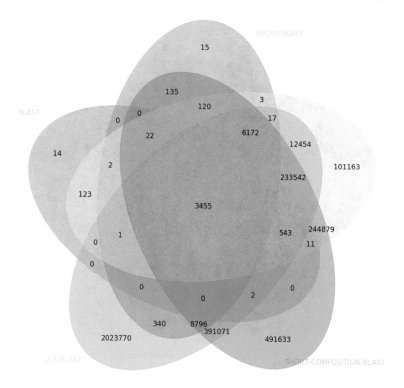

Fig. 3. Venn diagram presenting results from all modifications of BLAST.

number of similar pairs grows along with the length of the alignment due to higher scores.

3.1   Analyses of Differences Among Alignments

In order to show differences in alignments obtained with different methods we selected the P0C9V8 protein which was found by each method. Number of protein pairs with similar LCRs found by each method are presented in Table 2.

BLAST. Default BLAST found 5260 pairs of similar low complexity sequences. As presented in Fig. 6 these results contain a lot of mismatches. This number is really low in comparison to other modifications which is mainly caused by compositional based scoring matrix adjustment. Also, the results of BLAST contain similar pairs which are on the border of LCRs and HCRs. The most common amino acid found by BLAST method is serine (S).

SHORT-BLAST. BLAST with short sequences parameter turned on increased the number of similar LCR pairs in a result set to 1978. The most common amino acid in these results is glutamine (Q) which shows that glutamine is commonly

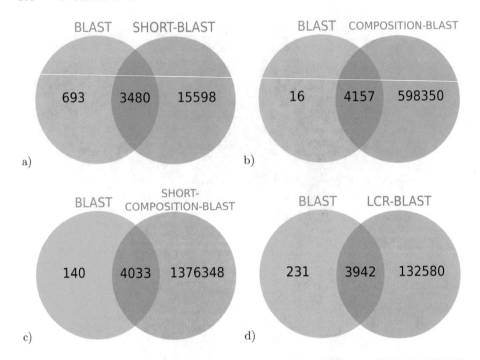

Fig. 4. Venn diagrams presenting overlap between (a) BLAST and SHORT-BLAST (b) BLAST and COMPOSITION-BLAST (c) BLAST and SHORT-COMPOSITION-BLAST (d) BLAST and LCR–BLAST.

found in short LCRs. Length of alignments generated by BLAST with short sequences parameter is presented in Fig. 5. BLAST with default parameters for short sequences found a lot of sequences near the length of 15 amino acids. However, for the analyzed protein P0C9V8 SHORT-BLAST found only two pairs of similar sequences. This is due to the fact that this version of BLAST uses PAM scoring matrix that has a greater penalty for mismatches. These alignments are shown in Fig. 7.

COMPOSITION-BLAST. Turning off composition based scoring matrix adjustment resulted in significant increase of similar LCR pairs because it is the main factor that masks LCRs in order to improve the search of evolutionary related proteins. This modification increased the number of similar LCR pairs to 922562. However, the results were inflicted with many mismatches. Example alignments are presented in Fig. 8.

In this version of BLAST, quality of the alignment is assessed by S value based on BLOSUM matrix which results in incorrect score given to particular types of alignments. For example, the 18 amino acid long homopolymer of glutamine has the e-value 6.07292e-05 with the score 90, whereas the homopolymer of asparagine of the same size has the e-value equal to 6.16153e-07 with score 108.

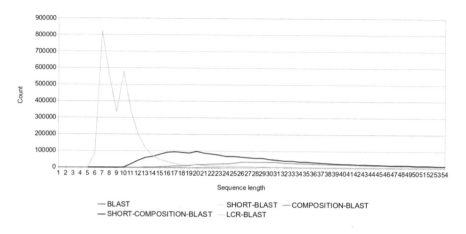

Fig. 5. Lengths of alignments found by each method.

Table 2. Number of sequences similar to the P0C9V8 protein for different BLAST parameter settings

Method	Number of similar LCR pairs
BLAST	5
SHORT-BLAST	2
COMPOSITION-BLAST	271
SHORT-COMPOSITION-BLAST	57
LCR-BLAST	5

```
P0C9V8   PCPPPKPCPPPKPCPPPKPCPPPKPCPPPKPCPPPKPCPPPKP
         P  P KP   P KP   P KP   P KP   P KP   P KP   P KP
P16053   PASPEKPRTPEKPASPEKPATPEKPRTPEKPATPEKPRSPEKP


P0C9V8   PCPPPKPCPPPKPCPPPKPCPPPKPCPPPKPCPPPKPCP
         P  +P P P+P P   +P    +P P P+P    P+P P P
Q2US45   PASRPTPKPQPLPQSQPATA-QPIPAPQPAAVPRPVPQP


P0C9V8   PNPCPPPKPCPPPKPCPPPKPCPPPKPCPPPKPCPPPKPCPPP
         P  PCPPPKPCPPPKPCPPPKPCPPPKPCPPPKPCPPPKPCPPP
P0C9V6   PKPCPPPKPCPPPKPCPPPKPCPPPKPCPPPKPCPPPKPCPPP
```

Fig. 6. Example alignments of similar LCRs found by BLAST

Intuitively these alignments should have the same score. That is because of the score values assigned to these amino acids in BLOSUM scoring matrix where the match of glutamine and asparagine score 5 and 6 respectively.

```
P0C9V8    PKPCPPPKPCPPPKPCPPPKPCPPPKPCPPP
          PKPCPPPKPCPPPKPCPPPKPCPP KPCPPP
P0C9V7    PKPCPPPKPCPPPKPCPPPKPCPPSKPCPPP
```

```
P0C9V8    PKPCPPPKPCPPPKPCPPPKPCPPPKPCPPPKPCPPPKPCSSP
          PKPCPPPKPCPPPKPCPPPKPCPPPKPCPPPKPCPPPKPC  P
P0C9V6    PKPCPPPKPCPPPKPCPPPKPCPPPKPCPPPKPCPPPKPCPPP
```

Fig. 7. Example alignments of similar LCRs found by SHORT-BLAST

```
PNPCPPPKPCPPPKPCPPPKPCPPPKPCPPPKPCPPPKPCPPP
P PCPPPKPCPPPKPCPPPKPCPPPKPCPPPKPCPPPKPCPPP
PKPCPPPKPCPPPKPCPPPKPCPPPKPCPPPKPCPPPKPCPPP
```

```
PKPCPPPKPCPPPKPCPPPKPCPPPKPCPPPKP
P+P  P  P+P  P  P+P  P  P+P  P  P+P  P  P+P
PQPQPRPQPQPQPQPQPQPQPQPQPRPQPQPQP
```

```
PCPPPKPCPPPKPCPPPKPCPPPKPCPPPKPCPPPKPCPPPKPC
P PPP  P  PPP  P  PPP  P  PPP  P  PPP  P  PPP+
PLPPPPPPPPPPPPPPPPPPPPPPP-PPPSPPPPPPPPPPPQ
```

Fig. 8. Example alignments of similar LCRs found by COMPOSITION-BLAST

SHORT-COMPOSITION-BLAST. BLAST with short sequence parameters turned on and disabled composition based adjustment of scoring matrix resulted in the highest number of similar pairs in comparison to the rest of the methods. However, when we analyzed the quality of alignments we realized that some of them include a lot of mismatches such as alignment of P0C9V8 and Q84ZL0 presented in Fig. 9.

LCR-BLAST. Applying additional modification to the BLAST such as identity matrix and mean score improved the quality of alignments. Example alignments are presented in Fig. 10. We accept up to 1.3 mutation per 6 amino acids. Using LCR-BLAST we found 3354495 pairs of similar LCRs. BLAST found similar LCRs in 231 pairs of proteins. However, in Fig. 6 the alignment between P0C9V8 and P16053 is far below assumed mean score threshold for LCR-BLAST. That is the reason why these pairs of proteins containing similar LCRs do belong to the LCR-BLAST set.

```
PFPLPNPCPPPKPCPPPKPCPPPKPCPPPKPCPPPKPC
PFPLP PCPPPKPCPPPKPCPPPKPCPPPKPC PPKPC
PFPLPKPCPPPKPCPPPKPCPPPKPCPPPKPCSPPKPC


PCPPPKPCPPPKPCPPP--KPCPPPKPCPPP--KPCPPPKPCPPP--KPCSSP
P PPP P PPP P PPP   P PPP P PPP   P PPP P PPP   P S P
PSPPPPPSPPPPP-PPPGARPGPPPP-PPPPGARPGPPPPP-PPPGGRP-SAP


PPPKPCPPPKPCPPPKPCPPPKPCPPPKPCPPPKPCPPPKP
PPP P PPP P PPP P PPP   PPP P PPP P PPP P
PPPPPPPPPPPPPPPPPPPPP---PPPPPSPPPPPPPPPPPP
```

Fig. 9. Example alignments of similar LCRs found by SHORT-COMPOSITION-BLAST

```
P0C9V8   PKPCPPPKPCPPPKPCPPPKPCPPPKPCP
         PK CPPPKPCPPPKPCPPPKPCPPPKPCP
P0C9V9   PKLCPPPKPCPPPKPCPPPKPCPPPKPCP


P0C9V8   PPKPCPPPKPCPPPKPCPPPKPCPPPKPCPPPKPCP
         PPKPC PPKPCPPPKPCPPPKPCPPPKPCPP KPCP
Q89501   PPKPCRPPKPCPPPKPCPPPKPCPPPKPCPPSKPCP


P0C9V8   PKPCPPPKPCPPPKPCPPPKPCPPPKPCPPP
         PKPCPPPKPCPPPKPCPPPKPCPP KPCPPP
P0C9V7   PKPCPPPKPCPPPKPCPPPKPCPPSKPCPPP
```

Fig. 10. Example alignments of similar LCRs found using LCR-BLAST

4   Conclusion

In this work, we compared our new LCR-BLAST method with BLAST using different parameter settings designed to analyse short sequences and sequences with bias in amino acids composition. We notice that BLAST with default parameters is not efficient enough for LCR analysis which is due to the fact that it is designed to search for evolutionary relationships between regular protein sequences.

Turning on default parameters for short sequences slightly increased the number of similar LCR pairs. These new sequences are mainly short. By turning off composition based statistics we can notice a huge increase in the number of similar pairs. That is because the score for these regions is not decreased due to scoring matrix recalculation. However, we can also observe a lot of mismatches in the result set.

Introduction of identity scoring matrix improved the search for similar sequences in the context of both homologous and analogous sequences. According to our best knowledge, this is the first approach to use a scoring matrix for analysis of LCRs. We proposed a very simple approach which treats equally all amino acids substitutions. In the future work we will investigate more complex models of scoring matrices suitable for LCRs analysis.

Finally, we introduced mean score statistics in order to allow the user to assess the similarity of the alignment by its composition instead of evolutionary relationship. This change allowed us to increase the number of similar pairs from 5261 found by BLAST to 3354496 found by LCR-BLAST. In contrast to the results of COMPOSITION-BLAST and SHORT-COMPOSITION-BLAST methods the quality of alignments was better as the provided score reflects the similarity of LCRs in a more realistic way.

Acknowledgements. The research was supported by Statutory Research of Institute of Informatics, Silesian University of Technology, Gliwice, Poland grant no. BK-204/RAU2/2019 (AG) and co-financed by the European Union through the European Social Fund, grant POWR.03.02.00-00-I029 (PJ).

References

1. Altschul, S.F., Gish, W., Miller, W., Myers, E.W., Lipman, D.J.: Basic local alignment search tool. J. Mol. Biol. **215**, 403–410 (1990)
2. Andrade, M.A., Perez-Iratxeta, C., Ponting, C.P.: Protein repeats: structures, functions, and evolution. J. Struct. Biol. **134**, 117–131 (2001)
3. Bhowmick, A., Brookes, D.H., Yost, S.R., Dyson, H.J., Forman-Kay, J.D., Gunter, D., Head-Gordon, M., Hura, G.L., Pande, V.S., Wemmer, D.E., Wright, P.E., Head-Gordon, T.: Finding our way in the dark proteome. J. Am. Chem. Soc. **138**(31), 9730–9742 (2016)
4. Consortium, T.U.: Uniprot: a worldwide hub of protein knowledge. Nucleic Acids Res. **17**, D506–D515 (2019)
5. Dayhoff, M.O., Schwartz, R.M.: Chapter 22: a model of evolutionary change in proteins. In: Atlas of Protein Sequence and Structure (1978)
6. Henikoff, S., Henikoff, J.G.: Amino acid substitution matrices from protein blocks. Proc. Natl. Acad. Sci. **89**(22), 10915–10919 (1992)
7. Karlin, S., Altschul, S.: Methods for assessing the statistical significance of molecular sequence features by using general scoring schemes. Proc. Natl. Acad. Sci. U.S.A. **87**(6), 2264–2268 (1990)
8. Kruger, R.: Illuminating the dark proteome. Cell (2016)
9. Marcotte, E.M., Pellegrini, M., Yeates, T.O., Eisenberg, D.: A census of protein repeats. J. Mol. Biol. **293**(1), 151–160 (1999)
10. Mier, P., Andrade-Navarro, M.A.: Glutamine codon usage and polyq evolution in primates depend on the q stretch length. Genome Biol. Evol. **10**, 816–825 (2018)
11. Smith, T.F., Waterman, M.S.: Identification of common molecular subsequences. J. Mol. Biol. **147**, 195–197 (1981)
12. Wootton, J.C., Federhen, S.: Statistics of local complexity in amino acid sequences and sequence databases. Comput. Chem. **17**(2), 149–163 (1993)

Risk Susceptibility of Brain Tumor Classification to Adversarial Attacks

Jai Kotia, Adit Kotwal$^{(\boxtimes)}$, and Rishika Bharti

Dwarkadas J. Sanghvi College of Engineering, Vile Parle, Mumbai, India
jaikotia10@gmail.com, adit.kotwal29@gmail.com, bhartirishika@gmail.com

Abstract. Discovery of adversarial attacks on deep neural networks, have exposed the vulnerabilities of these networks, wherein they often entirely fail to classify the attack generated images. While deep learning networks have generated promising results in performing brain tumor classification, there has been no analysis of their susceptibility to adversarial attacks. Vulnerability to adversarial attacks can render the deep neural networks useless for practical medical application. In this paper, a study has been performed to determine extent of white box adversarial attacks on convolutional neural networks used for brain tumor classification. Three different adversarial attacks are implemented on the network, namely Noise generated, Fast Gradient Sign, and Virtual Adversarial Training methods. The performance of the network under these attacks is analyzed and discussed. Furthermore, in the paper it is shown how these networks perform when trained on the adversarial attack generated images, which could be a possible solution to prevent the failure of the classification networks against adversarial attacks.

Keywords: Adversarial attacks · Brain tumor classification · Deep convolutional neural network

1 Introduction

Brain Tumor is a type of cancer that is observing a surge in the number of patients every year. It is an abnormal and often uncontrolled growth of cells, that takes up space within the cranial cavity. Brain tumors are of various types, but from a visual perspective, they all appear quite similar to the naked eye. For the past century, the classification of brain tumors has been carried out based largely on concepts of histogenesis. These concepts explain that tumors can be classified according to their microscopic similarities with different putative cells of origin and their presumed levels of differentiation [1]. These differences are so minute that they are difficult to perceive without technological intervention. Radiologists are challenged with the task of identifying which exact class does a brain tumor belong to and they need to do so with an extremely high rate of success. Medical imaging allows the radiologist to study the tumor closely and uncover finer details to identify their type. This process can take a lot of time and

© Springer Nature Switzerland AG 2020
A. Gruca et al. (Eds.): ICMMI 2019, AISC 1061, pp. 181–187, 2020.
https://doi.org/10.1007/978-3-030-31964-9_17

with an ever increasing number of patients, there is growing scope and need for automation of this classification. There is constantly evolving work to improve upon medical image processing and classification techniques. This is where deep neural networks are applied for classification of brain tumors. Based on recent works [2–5] it is observed that deep neural networks, provide very good accuracy in classifying brain tumor images. They outperform other conventional machine learning methods for classification and their positive results indicate that they could actually be used in practical set ups. Convolutional neural networks in particular, a special type of deep neural network designed specifically for image data, perform well in brain tumor classification.

The works mentioned, provide results of the deep neural networks used for classification. But at the time of writing, there has been no study on the susceptibility of these networks to adversarial attacks. Adversarial attacks can cause the network to fail while classifying images and provide wrongly classified results [6]. When introduced in medical image networks [7,8], these attacks can cause unforeseen harm if radiologists were to rely on them for outcome. Hence it is seen that although deep neural networks are highly successful in classifying brain tumor images, the susceptibility to adversarial attacks can render the network unsuccessful and dangerous [9].

Adversarial attacks introduce variations in the input image, such that the image appears almost the same as the original, but the neural network classifies it as a class different from the original [10]. These modifications are quantified by the $l\infty$ norm which deals with the maximum absolute change in the pixel. The underlying idea is that they fool the neural network by making small perturbations to the same image to output a totally different class. These attacks function by exploiting the pattern learned by a neural network in classifying training images. As the neural network is learning to classify input images, the adversarial attack aims to understand what image features is the classification output dependant upon. It then uses this information to make subtle changes to an input image and fool the neural network to classify it incorrectly.

2   Methodology

Brain Tumor Classification. For Brain Tumor Classification the network is fed with input data images and classified into three classes of tumors. The classification of brain tumor has been performed with various methods in previous works. Deep convolution neural networks have produced good results on the classification task as presented by Cheng et al. [2], Abiwinanda et al. [3] and Mohsen et al. [4].

Motivated by the success of the above mentioned works, the deep convolutional neural network has been used to classify the brain tumor images. The performance of the network is promising for use in practical medical applications. But before that, the susceptibility of the network to adversarial attacks is required to be tested. If the networks fails and misclassifies images after adversarial attack, then it may not be suitable for real world applications. This is a big drawback as the accuracy produced by the deep neural networks are very good and could be applied for automation of such tasks.

Adversarial Attack Approach. Three different adversarial attacks have been implemented on the network. The attacks performed in the experimentation are white box attacks. In white box attacks the attacker has access to the underlying training policy of the deep neural network such as the model structure, biases, weights and hyper parameters while it is under training. The three attacks used are Noise based, Fast Gradient Sign [11] and Virtual Adversarial Training [12] methods.

In essence, a noise-based attack can be implemented by adding noise to the image using a predefined rule. The Noise based attack is a weak attack that just picks a random point in the action space. The idea behind this is to add noise in the image using a rule. The rule involves sending it to the classifier after every step until the classifier makes a mistake. The attacker usually goes one step further to select the weakest point of minimum amplitude that causes misclassification in order to avoid being noticed. In the implementation, when random noise is added, data points are perturbed in a random direction of the input space [13]. The Fast Gradient Sign and the Virtual Adversarial Training methods are relatively stronger at fooling the deep neural networks.

The gradient based attacks in which the attacker modifies the image in the direction of the gradient of the loss function are known to be the most successful. The Fast Gradient Sign Method (FGSM) is one such approach which calculates the pixel-wide deviation in the gradient's direction. In the model, x represents the input to the neural network, y is the output and $J(\theta, x, y)$ represents the cost of training the neural network. The gradient is calculated using backpropagation. The cost function can be linearized at a certain value to obtain maximum perturbation using the formula [14],

$$x^{adv}_{FGSM} = x + \epsilon.sign(\nabla_x L(h(x), y_{true})) \tag{1}$$

where ϵ is a small number and ∇ is gradient of the cost function with respect to x. It is worthy to note that the aforementioned equation uses the sign of the elements of the gradient of the cost function with respect to input fed into the neural network to compute the perturbations.

The Virtual Adversarial Training (VAT) method uses virtual labels which are probabilistically generated and are used to calculate the direction of the adversarial modifications. Local Distributional Smoothness (LDS) is defined as the negative of the sensitivity of the model distribution $p(y|x, \theta)$ with respect to the perturbation of x. It is based upon two hyperparameters $\lambda > 0$ and $\epsilon > 0$ and the loss function is represented by LDS. The LDS is defined such that it can only be applied to models which have a smooth distribution with respect to x. Objective function in terms of them is given by the equation [12],

$$\frac{1}{N} \sum_{n=1}^{N} \log p(y^{(n)}|x^{(n)}, \theta) + \lambda \frac{1}{N} \sum_{n=1}^{N} LDS(x^{(n)}, \theta) \tag{2}$$

where $x^{(n)}$ represents the perturbation and $y^{(n)}$ represents the label.

Retraining on Adversarial Attack Generated Images. For testing if the convolutional neural network can learn to distinguish between adversarial images if allowed to train on the attacked images, it is trained again. But this time, it is fed in adversarial attack generated images as input to the network. Thus it is studied, how the model performs when given adversarial images to train on, which may help decrease the vulnerability of the network (Fig. 1).

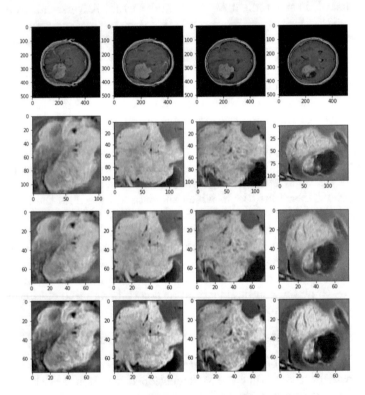

Fig. 1. The four rows of images show the data preprocessing as performed on four sample images from the data set belonging to different classes. The first row contains the original data images. The second row contains the cropped images of the tumor. In the third row the images have been scaled to a fixed size. Finally in the fourth row grayscaling is applied to the images, which is then fed as input images for the convolutional neural network.

3   The Data Set

The CE-MRI data set [16] was acquired from Nanfang Hospital, Guangzhou, China, and General Hospital, Tianjing Medical University, China, between 2005 to 2010. The data set used for this paper contains 3064 contrast enhanced images from 233 patients. The paper attempts to classify three different types of tumors: meningioma, glioma and pituitary tumor. While the tumor classes are distinct, to the human eye they appear quite similar and it is hard for radiologists to distinguish between them. The images of the tumor are taken from various angles, which makes the classification task harder.

Image Preprocessing. The images were preprocessed by first cropping the tumor region from the image. The tumor edge boundaries were provided in the data set, which had been manually plotted by experienced radiologists. The images were then scaled to obtain a fixed size for the input image data. Based on average values of width and height of the images, the images were scaled down to the size of 75×75. The image was grayscaled [15] to isolate spatial features and remove any inconsistencies in coloring.

4   Results and Discussion

For performing the experiments, all of the 3064 tumor images from the data set were used. The training and test data was split in the ratio 4:1, such that there were 2451 training samples and 613 test samples. The training was performed in two phases. First the convolutional neural network was subjected to adversarial attack and the accuracy of the model was tested. The second phase involved training the same convolutional neural network, but this time the adversarial attack generated images were included in the training data input.

Table 1. Performance of the convolutional neural network in classifying original and adversarial images for the three attacks.

Adversarial attack	Validation accuracy	Adversarial accuracy
Noise based	86.13%	62.15%
Fast Gradient Sign Method	81.24%	12.07%
Virtual Adversarial Training	87.93%	53.18%

Phase One. As seen in Table 1 the convolutional neural network produced good accuracy in classifying brain tumor images, of around 85% on validation data and up to 99% on training data. But the network failed miserably in classifying adversarial attack generated images. The accuracy significantly dropped as it attempted to classify the images generated by the various attacks. The network fared best when classifying Noise based attacks, as it is a weak method. Even so, an accuracy of only 62.15% is nowhere close to being applicable in practice. Against the Fast Gradient Sign Method it had very poor results where it failed almost entirely. The extremely low accuracy of 12.07% against the Fast Gradient Sign method shows how the network can dangerously be attacked to misclassify any input image. The network had a similar failure in classifying the Virtual

Adversarial Training generated images. These results demonstrate that the convolutional neural network is susceptible to adversarial attacks while classifying brain tumor images.

Table 2. Performance of the convolutional neural network in classifying original and adversarial images for the three attacks after training on adversarial attack generated images.

Adversarial attack	Validation accuracy	Adversarial accuracy
Noise based	88.58%	86.79%
Fast Gradient Sign Method	89.23%	89.07%
Virtual Adversarial Training	89.40%	75.20%

Phase Two. From Table 2 it can be observed that the convolutional neural network, upon training on adversarial attack generated images, produces far better results in classification. The accuracy of 89.23% for the Fast Gradient Sign method, for which the network had performed the worst in *Phase one*, is an encouraging result. For both the Noise based and Fast Gradient Sign method, the validation accuracy and the accuracy after training upon adversarial attack generated images is almost the same, which means the network has successfully overcome the damage caused by the attacks. The accuracy on the Virtual Adversarial Training is lesser at 75.20% suggesting that even after training on adversarial attack generated images, it hampers the network classification accuracy.

The accuracy after training on adversarial attack generated images, suggests that if the convolutional neural network is trained on these images then the vulnerability to these attacks considerably reduces. The network becomes more robust and its accuracy is almost the same on the adversarial attack generated image as its accuracy on the original images. This is a promising result as it indicates that the attacks can themselves be used to improve the security of the convolutional neural network.

5   Conclusion

In this paper, the susceptibility of a deep convolutional neural network was thoroughly tested against three different adversarial attacks. It was observed that the network failed in classifying adversarial attack generated images for the Fast Gradient Sign and Virtual Adversarial Training methods, although it showed relatively better resilience for the weaker Noise based attack. It was also observed that the network produces considerably good results in classification after training on adversarial attack generated images. These results show that the network can be trained to be equally adept at classifying original and adversarial attack generated images. Hence to increase immunity against adversarial attacks, these deep neural networks can be subjected to adversarial attack training before being deployed for practical use. This solution ensures that the deep neural network is less at risk against harmful adversarial attacks and maintains its accuracy while classifying brain tumor images. For future work, the susceptibility of these networks should be tested against black box attacks as the attacker may not always have access to the network architecture. Identifying if there is a bias developed within the model for classifying the input images can also be explored in future works.

References

1. Louis, D., et al.: The 2016 World Health Organization Classification of Tumors of the Central Nervous System: A Summary. Springer, Heidelberg (2016)
2. Cheng, J., Huang, W., et al.: Enhanced performance of brain tumor classification via tumor region augmentation and partition. PLoS ONE **10**(12), e0144479 (2015)
3. Abiwinanda, N., Hanif, M., et al.: Brain tumor classification using convolutional neural network. In: IFMBE, vol. 68/1 (2018)
4. Mohsen, H., El-Dahshan, E.-S.A., et al.: Classification using deep learning neural networks for brain tumors. Faculty Comput. Inform. J. **3**(1), 68–71 (2018)
5. Seetha, J., Raja, S.S.: Brain tumor classification using convolutional neural networks. Biomed. Pharmacol. J. **11**(3), 1457–1461 (2018)
6. Kurakin, A., Goodfellow, I., Bengio, S.: Adversarial examples in the physical world. arXiv:1607.02533 [cs.CV] (2016)
7. Finlayson, S.G., Chung, H.W., et al.: Adversarial attacks against medical deep learning systems. arXiv:1804.05296v3 [cs.CR]
8. Finlayson, S.G., Bowers, J.D., et al.: Adversarial attacks on medical machine learning. Science **363**(6433), 1287–1289 (2019). https://doi.org/10.1126/science.aaw4399
9. Papernot, N., McDaniel, P., et al.: The limitations of deep learning in adversarial settings. CoRR, abs/1511.07528 (2015). arXiv:1511.07528 [cs.CR]
10. Akhtar, N., Mian, A.: Threat of adversarial attacks on deep learning in computer vision: a survey. IEEE Access **6**, 14410–14430 (2018). https://doi.org/10.1109/ACCESS.2018.2807385
11. Zuo, C.: Regularization effect of fast gradient sign method and its generalization (2018)
12. Miyato, T., et al.: Distributional smoothing with virtual adversarial training. In: International Conference on Learning Representations (ICLR) (2016)
13. Fawzi, A., Moosavi-Dezfooli, S., Frossard, P.: Robustness of classifiers: from adversarial to random noise. In: Neural Information Processing Systems (NIPS) (2016)
14. Tramer, F., Kurakin, A., et al.: Ensemble adversarial training: attacks and defenses, arXiv preprint arXiv:1705.07204 (2017)
15. Xie, Y., Richmond, D.: Pre-training on grayscale ImageNet improves medical image classification. In: Leal-Taixé, L., Roth, S. (eds.) Computer Vision - ECCV 2018 Workshops. ECCV. LNCS, vol. 11134. Springer, Cham (2018)
16. Cheng, J.: Brain tumor dataset. Figshare. Dataset (2017). https://doi.org/10.6084/m9.figshare.1512427.v5

Prediction of Drug Potency and Latent Relation Analysis in Precision Cancer Treatment

Jai Kotia[✉], Rishika Bharti, and Adit Kotwal

Dwarkadas J. Sanghvi College of Engineering, Vile Parle, Mumbai, India
jaikotia10@gmail.com, bhartirishika@gmail.com, adit.kotwal29@gmail.com

Abstract. As cancer treatments are gaining momentum, in a bid to improve drug potency, doctors are looking towards precision cancer medicine. Here the drug prescriptions are tailored to the patients gene changes. In this paper, the aim is to automate the task of drug selection, by predicting the clinical outcome of using a particular drug on a combination of the patients gene, variant and cancer disease type. While the main idea behind precision cancer treatment is to identify drugs suitable to each patients unique case, it is justifiable for to assume that there exists a predictive pattern in these prescriptions. We propose to implement this prediction using three machine learning models, the Support Vector Machine, the Random Forest Classifier and the Deep Neural Network. The models yielded promising results of over 90% accuracy and over 95% ROC-AUC score. This positive outcome affirms the assumption that there exists a predictive pattern in precision treatments, that could be extrapolated to help automate such tasks. We further analyzed the data set and identified latent relations between drug, cancer disease, target gene and gene variant. This exploration uncovered some significant patterns where it can be observed how a particular drug has had successful results in treating a particular cancer and targeting specific gene variants.

Keywords: Precision cancer treatment · Machine learning · Latent relation analysis

1 Introduction

Of the various diseases that afflict the modern world, cancer is one of the most predominant [1]. More than 100 different types of cancers have been identified till date that are seen in residents of different parts of the globe and of different age groups. A characteristic trait among the cancers which makes it the second leading cause of death worldwide is its inherent variability. Cancer is caused by uncontrolled cell growth and malignant tumor formation which can be engendered by a wide variety of factors. Conventional treatments designed as an attempt to curb the disease fail due to their homogeneity, that is, standard

© Springer Nature Switzerland AG 2020
A. Gruca et al. (Eds.): ICMMI 2019, AISC 1061, pp. 188–195, 2020.
https://doi.org/10.1007/978-3-030-31964-9_18

regimens like radiation, immunotherapy, surgery and chemotherapy prove their effectiveness in only a fraction of the patients because the fail to take into consideration the broad spectrum of underlying causes and individual responses of each patient to the treatment. In fact, the generic drugs used for cancer treatment are noted to be ineffective in an astonishing 75% of the cases [2]. This is where precision treatment proves its worth.

Precision treatment is a type of medical care in which personalized medicine dozes and treatment plans are tailored for individual patients after a thorough understanding of the genetic mutations involved, how the tumor is growing and spreading within the body and how the patient is responding to treatments [3]. Additionally, other personal factors such as type, stage, location of the cancer, patient's age and overall health are also stressed upon. Although drug responses can vary individually based on age, sex, disease, or drug interactions, the genetic factors also influence the efficacy of a drug and the possibility of an adverse reaction [4–7]. The ideas revolving around precision treatment are not new to health care industry, yet so far the field had not been opened up to the masses. Only recently, with the advances in technology has precision medicine been on the path to achieve its true potential [8].

Machine learning and associated fields have helped payers, patients and providers recognize the remarkable power within this field [9,10]. It has also helped to reduce the costs associated with traditional *trial-and-error* treatments. The abundance of patient data regarding their genetic makeup and medical history that has been gathered in recent years provides a great source to establish machine learning models that can automate treatment prescriptions. Machine learning models have the ability to emulate or even improve upon human performance if given enough data to learn from [11].

2   Methodology

The task is to ultimately classify if a drug will produce a response from the target gene variant or will it show resistance for a given set of patient gene, variant and cancer disease type combination. For this three machine learning models were selected, namely, Support Vector Machine, Random Forest Classifier and the Deep Neural Network. The selection of models was based on their general success in classification tasks.

Following are the details of the three machine learning models used for performing the prediction of clinical outcome:

i *Support Vector Machine*

First the Support Vector Machine classifier was used for predictions. It is a popular machine learning model for classification tasks. The Support Vector Machine classifier was configured to use the *rbf* kernel, or the Radial Basis Function kernel.

ii *Random Forest Classifier*

Next the Random Forest Classifier was used, which is another popular machine learning model. It is an ensemble model that fits multiple decision tree classifiers to improve the classification accuracy. It works on a voting based system to select the best solution and prevents over fitting. An ensemble of 500 decision trees (estimators) were used for the Random Forest model.

iii *Deep Neural Network*

Finally, a Deep Neural Network is implemented on the data set. Recent works suggest that neural networks have been outperforming most of the other conventional models for classification. Motivated by this, a Deep Neural Network is implemented that has a depth of three layers. The three dense neuron layers are preceded by a dropout layer that avoids overfitting on the training data and use the ReLu (rectified linear unit) activation layer. The output layer uses the sigmoid activation layer. Finally the mode is compiled using the Adam optimizer and the binary cross entropy loss function which aids in the classification task.

3   The Data Set

The Clinical Interpretation of Variants in Cancer data [12] was used for this research, which is a database of precision cancer treatment records. The entries submitted by experts have been made available publicly for research and inference purposes.

Selection of data was made pertaining to our research goals. The following columns were included: gene, variant, drugs, disease, clinical significance, rating and evidence Level. The data contained 359 types of drugs, 311 types of genes, 829 variants and 206 types of cancer diseases. Majority of the clinical evidence available was based on predictive analysis.

For implementation, first a minimum rating of 3 stars (on a scale of 1 to 5, with 5 being the best) was assigned and a minimum evidence level of C (on a scale of A to E, with A being the best), so that only sound and well documented evidences were used for the predictions.

For the later part of predictions, a minimum frequency count of 5 was used for the model predictions. This means that drugs, genes, variants and diseases were selected constrained by a minimum frequency count. The minimum frequency count indicates how many times it appeared in the data. It was done remove items with examples that are too few to make any reasonable inference. For this data though, the minimum rating and evidence levels were removed so that there is more data to work with.

For data pre-processing one hot encoding on the input column values (gene, variant, drug and disease) was employed and dropped any rows with missing column values. The prediction output column (clinical outcome) was label encoded. The data set was split in the ratio of 3:1 for training and test samples.

4   Results and Discussion

The results as shown in Table 1 are encouraging and indicate that the assumption about some predictive pattern in precision cancer treatment was correct. As expected the Deep Neural Network model performed better than the other models.

Table 1. Prediction results of clinical outcome trained on Support Vector Machine (SVM), Random Forest Classifier (RFC) and Deep Neural Network (DNN)

Condition	Metric	SVM	RFC	DNN
Minimum rating level = 3	Test Accuracy	65.06	**78.92**	**78.92**
Minimum evidence level = C	ROC-AUC Score	50.00	74.22	**85.96**
No rating or evidence level limit	Test Accuracy	63.96	**86.94**	84.23
	ROC-AUC Score	50.00	83.65	**89.74**
No rating or evidence level limit	Test Accuracy	**93.14**	92.16	92.16
Minimum frequency = 5	ROC-AUC Score	93.29	92.13	**95.20**

The Support Vector Machine fared poorly in the first two prediction conditions, with and without minimum rating and evidence levels. The ROC-AUC score of 50.00% is the worst possible outcome and rightly so the Support Vector Machine gave unconvincing prediction accuracy for the first two conditions. But surprisingly it had a good accuracy in predicting clinical outcomes with a minimum frequency. This can be attributed to the fact that there are better relations between the input data columns with a minimum frequency count.

The Random Forest Classifier had decent results in all prediction conditions. Its accuracy was, in general, better than the Support Vector Machine but worse than the Deep Neural Network. The Random Forest Classifier benefits from the fact that it takes the best prediction of several decision trees (500 in this case).

The best performance was displayed by the Deep Neural Network. It consistently had a good ROC-AUC score complimented with good accuracy in the last two prediction conditions. The Deep Neural Network gave very promising results in predicting clinical outcomes on data with a minimum frequency count. The ROC-AUC score of 95.20% is excellent and provides scope for use in practical medical applications.

It can be inferred from Table 1 that the machine learning models can make very accurate predictions if a minimum frequency count is imposed on the input data. This suggests that there exists some relations between the gene, variant, cancer disease type and the drug used to treat the cancer. This information can be used to propagate scientific research in exploring such relations. There is indication that if more data was available, the accuracy for the first two prediction conditions can also be improved. This statement is made on the basis that there was very less data available after imposing the conditions. Thus there is strong potential of using machine learning models which can aid doctors in automating precision cancer treatment prescriptions.

5   Analysis of Latent Relations Between Gene, Variant, Drug and Disease

A gene is a physical and functional unit of heredity that passes genetic information between parent and offspring. The traits expressed by a cell, how it responds to atypical conditions, how it divides, what it must become (in the case of a stem cell) and how long it lives, have various pathways, all of which are concerned with the genes. As cancer is the uncontrolled growth of certain cells, it greatly influenced by genes. A genetic mutation, which is an alteration in the normal sequence of the genes is one of the primary causes of cancer. The modified arrangement in the genes results in abnormal behaviour of particular cells which can pervade into surrounding normal cells and result in a cancerous growth. It is worth to note that though cancers that hereditary genetic mutations may be a rare cause of cancer, but cancer development in the body is highly influenced by these mutations. Cancer therapies are designed keeping in mind the common genes that affect cancer growth irrespective of their place of origin in the body. Hence the aim is to understand and identify potential relations between these cancer causing gene variants and the drugs which successfully evoke response from them. It is also explored if specific gene mutations are the dominant cause of a particular cancer type.

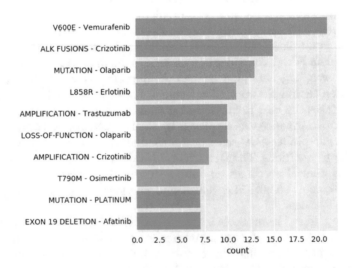

Fig. 1. Variant-drug relation frequency mapping

For forming the latent relation mappings in all three of the figures (Figs. 1, 2 and 3), first the data set was sliced to include only those entries where the drug used has evoked a sensitivity or response from the gene for a particular cancer disease type. Then the ten most common pairing in the figures were included, such that they show the combinations of variant-drug, gene-disease and drug-disease that are most prominent. The x-axis of the figures mark the frequency count, while the y-axis labels the pairings.

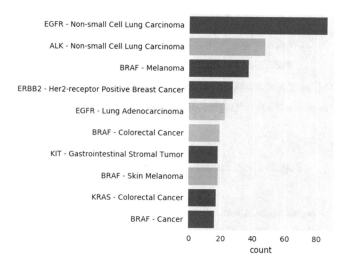

Fig. 2. Gene-disease relation frequency mapping

The variant-drug relation mapping has been shown in Fig. 1. Here it can be observed how often has a drug been used in an attempt to successfully target a particular gene variant. This relation mapping can be extremely useful for doctors in deciding which drug displays potency in targeting a gene variant. As seen between the variant V600E of the gene BRAF and the Vemurafenib drug, they have a striking occurrence frequency of over 20 for successfully evoking a response. It has been found that this latent relation has also been confirmed experimentally. There was an improved survival rate of patients with Melanoma when Vemurafenib was tried on a subgroup of Melanoma patients with a mutation in the BRAF V600E gene [13]. Thus further such relations between gene variant and drugs can be explored, which can help treat cancer in patients.

In Fig. 2 the relation between gene and the cancer disease type can be studied. With seven gene-disease pairings occurring more than 10 times, it shows that some cancer diseases are predominantly caused by specific genes. Especially in the case of EGFR - Non-small Cell Lung Carcinoma, occurring over 50 times, a strong relation between the pair can be observed. This information can help while diagnosis and in identifying the cancer causing genes. In precision medicine, knowing the gene causing the cancer can help save costs by eliminating use of futile medicines to which the patient may not respond.

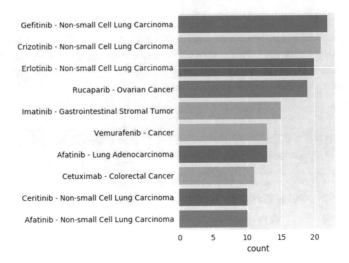

Fig. 3. Drug-disease relation frequency mapping

Figure 3 presents the latent relations between drug and the cancer disease type. While due to a large amount of entries about Non-small Cell Lung Carcinoma it dominates this chart, useful relations of Rucaparib for Ovarian Cancer and Nerantinib for Breast Cancer can still be observed. These findings can significantly assist doctors in narrowing down to a specific set of drugs that have proven to be successful in treating the particular cancer disease type.

Such analysis concurs with systems like EDGAR (Extraction of Drug, Gene and Relation) which is a natural language processing system that extracts relevant information regarding cancer genes and drugs in use from biomedical literature [14].

6   Conclusion

We have proposed a machine learning approach to predict the clinical outcome of precision cancer treatment. The results were promising and supported the assumption that there exists certain predictive relations between gene, variant, drug and cancer disease. This discovery can considerably help in automating precision cancer treatments, especially by using Deep Neural Networks, that presented encouraging accuracy and ROC-AUC scores.

We also analyzed latent relations between gene, variant, drug and cancer disease type. This presented potentially useful findings that showed frequent associations between variant-drug, gene-disease and drug-disease pairings. This information could be used by doctors to assist in narrowing down potential causes and drugs prescriptions for precision cancer treatment.

While these findings have given optimistic results, the works in precision cancer treatment could be improved if more functional information is gathered, during cancer therapies for example, instead of relying only on initial conditions [15]. Together with machine learning models, if equipped with more extensive data, much better results can be expected and eventually automation in precision cancer treatment could be achieved.

References

1. Siegel, R.L., Miller, K.D., Jemal, A.: Cancer statistics, 2019. CA Cancer J. Clin. **2019**(69), 7–34 (2019)
2. The Personalized Medicine Report 2017: Opportunity, Challenges and the Future. Personalized Medicine Coalition (2017)
3. Ashley, E.A.: The precision medicine initiative: a new national effort. JAMA **313**(21), 2119–2120 (2015). https://doi.org/10.1001/jama.2015.3595
4. Richard, W.: Inheritance and Drug Response. N Engl. J. Med. **348**, 529–537 (2003). https://doi.org/10.1056/NEJMra020021
5. Kalow, W.: Pharmacogenetics: Heredity and the Response to Drugs. W.B. Saunders, Philadelphia (1962)
6. Price Evans, D.A.: Genetic Factors in Drug Therapy: Clinical and Molecular Pharmacogenetics. Cambridge University Press, Cambridge (1993)
7. Weber, W.W.: Pharmacogenetics. Oxford University Press, New York (1993)
8. Friedman, A.A., Letai, A., Fisher, D.E., Flaherty, K.T.: Precision medicine for cancer with next-generation functional diagnostics. Nature Rev. Cancer **15**(12), 747–756 (2015)
9. Ding, M.Q., Chen, L., Cooper, G.F., Young, J.D., Lu, X.: Precision oncology beyond targeted therapy: combining omics data with machine learning matches the majority of cancer cells to effective therapeutics. Mol. Cancer Res. **16**(2), 269–278 (2018). https://doi.org/10.1158/1541-7786.MCR-17-0378
10. Daemen, A., et al.: Modeling precision treatment of breast cancer. Genome Biol. **14**, R110 (2013)
11. Stallkamp, J., Schlipsing, M., Salmen, J., Igel, C.: Man vs. computer: benchmarking machine learning algorithms for traffic sign recognition. Neural Networks **32**, 323–332 (2012)
12. Griffith, M., et al.: CIViC is a community knowledgebase for expert crowdsourcing the clinical interpretation of variants in cancer. Nat. Genet. **49**, 170–174 (2017)
13. Rindflesch, T.C., et al.: EDGAR: extraction of drugs, genes and relations from the biomedical literature. Biocomputing **2000**, 517–528 (1999)
14. Chapman, P.B., et al.: Improved survival with vemurafenib in melanoma with BRAF V600E mutation. Natl. Engl. J. Med. **364**, 2507–2516 (2011). https://doi.org/10.1056/NEJMoa1103782
15. Letai, A.: Functional precision cancer medicine–moving beyond pure genomics. Nature Med. **23**, 1028 (2017)

Predictions of Age and Mood Based on Changes in Saccades Parameters

Albert Sledzianowski[✉]

Polish-Japanese Academy of Information Technology, 02-008 Warsaw, Poland
asledzianowski@gmail.com

Abstract. We have measured saccades of 12 subjects in age below 30 (younger adults) and over 60 (older adults) during musical sessions with energetic and relaxing music. We wanted to check if saccade parameters are sensitive to emotions caused by music and if so, check if there ability to distinguish and classify those parameters to one of defined group of Age (young/old) or group of Mood (energetic/relaxed) induced by the music. We used combination of different types of filters and classifiers from WEKA [5] in search of possible correlations between saccade parameters and characteristics of respondents. Results showed statistical changes between age groups in the latency (23.6% of difference) and in the pupils size (16,6% of difference), both found significant ($P < 0.0001$). In case of Mood, results showed changes in the group of younger adults in the latency ($P = 0.4532$), the amplitude ($P = 0.0001$) and for the average velocity ($P = 0.0048$). Prediction of age group showed the accuracy of 91.4%, in case of Mood groups 97%. For both types of groups, best predictions were obtained by the Random Forest and the Multilayer Perceptron.

Keywords: Eye moves · Saccades · Eye tracking · Eye move computations · Brain computations · Human features · Emotions · Moods · Age · Machine detection and classification

1 Introduction

We investigated possible changes in parameters of saccades, caused by emotions induced by music listening, among people who are different in the age. Music has a fundamental effect on humans. It was shown that during motor activity, music optimizes arousal, facilitates task-relevant imagery and improves performance in simple and repetitive motor tasks, like saccadic eye movements [12]. Music also has significant effects on Autonomic Nervous System (ANS) especially on heart rate, respiration rate, blood pressure and also on pupillary response controlled by ANS. In states of relaxation, limbic moves (thus also eye moves), theoretically should be slower and less rapid, while in state of arousal, faster and more dynamic. Neutral music significantly reduces eye movement activity, when they listened to music subjects exhibit longer fixations, more blinks and fewer saccades [18]. One of the basic types of eye movements is the saccade which can be

© Springer Nature Switzerland AG 2020
A. Gruca et al. (Eds.): ICMMI 2019, AISC 1061, pp. 196–205, 2020.
https://doi.org/10.1007/978-3-030-31964-9_19

described as rapid, ballistic movement of the eyes that rapidly change the point of fixation [4]. Saccades are varied in terms of visual angle, from small saccades made when we read, to longer movements when we look around ourselves. The more points are deviated from each other, the shorter is the fixation and saccades are more dispersed [6]. Saccades occur involuntarily whenever the eyes are open, even they are fixated on a target, but can be also elicited voluntarily [4]. Human eyes make saccades when watching available field of view, fixing on interesting elements of the surroundings. The source of this phenomenon comes from the fovea, which receives input from only a small portion of the visual field. In order to view a complex scene, the fovea must be moved to new fixation locations very often [9]. The speed of saccade cannot be controlled - the eyes move as fast as they are he highest speed at around 15° [1]. Saccades are very common in the functioning of the oculomotor system, occurs up to 173,000 times per day [17]. Saccade generation and triggering seems to be very sensitive as many areas of the brain are involved in this process. The results of studies on selective effect of age on oculomotor performance states, that in neurologically normal individuals, senility is serving to slightly increase saccade latencies [15]. Older adults have longer latencies in a variety of motor tasks (i.e. aimed hand movements) [9,15]. The general slowing hypothesis, suggests that slowing of information processing is increasing with age and it is an overall slowdown, not connected with any of particular tasks [9]. It seems to be true, as specific task may also be affected by other age related factors. In case of saccade trials there is an evidence indicating that as people becomes older, inhibition of inappropriate response becomes less effective, as well as suppression of irrelevant information. In addition, depression or anxiety both of which are common in older people and increasing with age, may also interact and impair or alter saccadic task performance [15]. However it also seems that older adults produce saccadic eye movements in fundamentally the same way as younger adults. It may suggest that mechanism which triggers and controls the saccade may not be degraded with progressing age [9]. In terms of pupil size, few researches states a significant difference between age groups. The rate of change of pupil diameter decreases with age from 0.043 mm per year at the lowest luminosity level to 0.015 mm per year at the highest [3], so the average pupil size of subjects between 20–30 years old is $\sim$ 6.9–6.5 mm, while pupil size of adults between 60–70 is $\sim$ 5.1–4.8 mm [10]. In case of pupil size, it has been found that larger dilations were predicted for emotions connected to arousal or tension in case of musical excerpts rated as inducers of these emotional states. Overall, findings suggests a complex interplay between different influences on pupillary responses to the music [2]. The experiment described in this article, was conducted to determine possibilities of prediction of age and mood measurements from saccadic parameters, in context of emotions induced by music listening - which is not well studied and understood yet.

2   Methods of the Experiment

12 subjects, different gender and aged 23 to 67 divided into two groups of young (six subjects in age between 23–30, mean 25.75, ± 1.47 SD) and old adults (six

subjects in age between 60–67, mean 64.75, ± 1.47 SD). None of participants indicated any illness, nor take any medicine. Each of participant declared before the study, that music was capable of exerting an emotional effect on him - relaxing and stimulating. It can be said, that only people who declared emotional sensitivity to music took part in the experiment. Subject's eyes movement during saccade sessions were measured by eye tracking device The Eye Tribe ETI1000 (60 Hz). During each session all subjects viewed IBM Think Vision 6636 15" LCD square display (1024 × 768 px) from distance of 60 cm in the sitting position.

During the experiment visual elements were displayed on the screen and managed by dedicated application. At the beginning of each trial, subject was viewing a fixation point - the green ellipse located in the middle of the screen and that was the primary position of the gaze for each saccade trial. To initiate saccade measurement process, subject had to keep eyes on the fixation point for randomly selected time period between 1000 and 2000 ms. Eye movements were measured in response to fixation and target points were switched sequentially on and off. The target always appeared in the same horizontal location of the fixation point and has the same shape of the ellipse. Schema on Fig. 1 presents the model of the saccade trial.

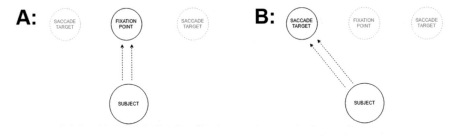

Fig. 1. Model of the saccade trial

Subjects were asked to perform saccades from fixation point to the appearing target. Before each saccade attempt, application set a random break period (0.5, 1 or 1.5 s) performed to help release gaze tensions and also to avoid eye to start moving rhythmically. Thus, subject never knew exactly when another trial starts. When measurements started, fixation point disappeared. In the same moment, target of the saccade appeared randomly on one of sides of the screen. This "Gap Effect" measurement model introduced by Saslow (1967), assumes that the removal of a visual fixation point shortly before the appearance of a peripheral target results in shorter latencies for saccades directed at the target, than when the fixation point remained on. As stated previously, according to other researches [11,12], latency is the main difference in saccade parameters between different age groups and older people are less able to suppress saccades during active fixation and have reduced effect of fixation point offset relative to younger people. This effect of reduced fixation inhibition is beneficial for older people in terms of decreased length of latency [15]. The Gap model shows latency adjusted by this phenomenon, thus shows reduced differences in response time between both age groups.

Before the experiment, subjects had to choose two pieces of music they knew and they could say that it most affects their emotions. It was told, that first piece should make subject energetically stimulated, the second one should make subject calmer and put him in quiet or touched mood. Subject could find and play any pieces of music. Participants choose their own pieces, because perception of music is very subjective and it is difficult to find pieces which could influence every participant. Musical piece was listened to on closed headphones before sessions of Energetic and Relaxed mood. It has been also determined that the respondents will choose music they know and like, as pupillary responses should modulate less strongly when track is particularly liked [2], thus should reduce differences between different mood groups.

Experiment was conducted in different 3 sessions of saccades recording, for "music off", energetic and relaxing music. Each session type was expressed in numeric value related to the assumption of object's stimulation: 0 for "music off", −5 for relaxing, 5 for energetic music. Each session contained preparations and/or music listening, device calibration and saccade measurements procedure. Session started only when device was perfectly calibrated, which was determined programmatically. Music was played before, not during the saccade measurements to avoid interactions with Autonomic Nervous System [8]. After each session subject's mood was noted in questionnaire. Mood was measured by object himself in scale between −5 and 5, where 0 is the neutral state, −5 is the most calm and 5 is the most excited. Each of sessions contained 3 sequences of 10 saccade tries, each session lasted about 15 min. Subject himself decided when to start another sequence of saccade measurements, so he could rest between them. Subject also decided, when to stop listening to music. It was said that it should be a moment when he felt the music had a profound effect on him.

3   Computational Basis

The output from the eye-tracker contains raw data of eye properties and coordinates measured in frequency of 60 Hz. Data log of each saccade was analyzed by application which calculated saccades latency, duration, amplitude, average velocity, peak velocity and average pupil size.

The latency was measured as period between fixation point disappearance and start of the saccade. Start of the saccade was set as the moment when gaze velocity (deg/sec) reached over the threshold velocity determined for particular record. Duration of the saccade was determined as the period, starting when gaze velocity (deg/sec) was over velocity threshold and finish when fell down below it. Threshold velocity was calculated from all subsequent frames of the record (including latency period) by dividing the maximum speed by the average speed. Additionally application detected and recorded the moment when subject's gaze focus left the fixation point and when it reached the target (which served as additional validation). The amplitude was calculated based on the visual angle.

The visual angle of the image on the retina can be calculated by measuring the angle subtended by the object outside of the eye [11]. Thus, the visual angle can be calculated as the arctangent of tracked distance of subject's eyes position from fixation point (b) to the target (c) divided by distance (d) of subject's eyes to the display (the tangent). The distance from fixation point to the target was calculated from screen coordinates (X, Y) obtained from each subsequent eye-tracker frames in period of saccade duration. Saccade record usually consisted 3–5 frames. Each subsequent distance between frame's coordinates was calculated as Euclidean Distance converted from pixels to centimeters. The distance between subject's eyes and the screen (d) was measured manually before each session. Average angular velocity expressed in degrees per second was calculated as the division, where amplitude (visual angle) obtained from saccade on-screen distance and distance between subject's eyes and the screen is the dividend and count of frames collected from the eye-tracker during period of duration is the divider, multiplied by eye-tracker frequency of 60 Hz. Peak angular velocity was counted as maximum value from velocity values, calculated and collected for every eye-tracked frame in the period of saccade duration. Pupil size was counted as the average of pupil size values from all frames collected by the eye tracker in the period of saccade duration.

Complete class of particular saccade contained following properties (Table 1). Properties: ID, Age, Sex describe subject, properties: Music, Mood describe session and properties: SccLat (Latency in ms), Dur (Duration in ms), Amp (Amplitude in deg), AvgVel (Average Velocity in deg/sec), AvgPupSize (Average Pupil Size in mm) describe saccade parameters.

Table 1. Saccade trial data row.

Id	Age	Sex	Music	Mood	Lat	Dur	Amp	Avg Vel	Peek Vel	Avg PupSize
Id	Age	M, F	−5, 0, 5	−5, −4, −3, −2, −1 0, 1, 2, 3, 4, 5	Lat	Dur	Amp	Avg Vel	Peek Vel	Avg PupSize

Each of 3 sessions was conducted in 3 sequences of 10 saccade trials. In order to find regularities in calculated data, WEKA 3.6.13 was used for attribute selection, evaluation of means and classifications.

4   Results

We analyzed the results for attributes placed in the dataset except the attribute "Music". This experiment concerned mood influence on saccade parameters in different age groups and attribute "Music" was the contractual parameter defining session type. True feelings after listening to the musical piece, were indicated by subject himself in value of attribute Mood. Many times music did not have the desired effect and subject responded with contradictory value for "Mood", thus attribute "Music" was not taken into account during the dataset analysis.

First, we analyzed dataset contained Age property and saccade results and all subjects from both age groups. After attributes discretization, 2 different age groups of younger {<30 } and older adults {>60 } were selected for comparison of saccade parameters. Dataset assembled from both groups, was then filtered for attribute selection using WEKA Exhaustive Search and the Cfs Subset Evaluator [5], resulting in selection of attributes: Age, Latency, Amplitude and Pupil Size. Entire dataset contained 311 instances. Statistics are presented in Table 2.

Table 2. Comparison of mean values for age groups.

Age groups	Latency (ms)	Amplitude (deg)	Pupil size (mm)
Younger adults (<30)	249.69 ± 60.1 (SD)	5.12 ± 0.8 (SD)	6.9 ± 1.2 (SD)
Older adults (>60)	326.70 ± 89.7 (SD)	5.05 ± 0.7 (SD)	5.9 ± 1.4 (SD)
Differences (%)	23.6	1.4	16,6
df	10	10	10
Standard error of difference	8.673	0.085	0.148
P	<0.0001	0.4129	<0.0001

Attributes Latency and Pupil Size have been found statistically significant. Size of human's pupil varies in range of 2–8 mm, so 1 mm difference is equal to the change of 16.6% in this scale. According to the results, saccades with average pupil size below 4.79 mm or saccades with latency over 265 ms might be assigned to the set of outputs of people over 60 years old.

Attribute Mood represents true feelings of subject after listening to the musical piece. Value of this attribute was indicated each time after music listening, by subject himself, in scale of −5 and 5, where −5 is the most relaxed and 5 the most energetic value of mood. There was 3 sessions types: without music (0), with energetic (5) and with relaxing music (−5). Mood value was given by subject himself at the end of each type of the session. Value was noted in questionnaire attached to each of session type records of particular subject. For some of the records, musical piece haven't influenced the subject, so mood value was not corresponding with session/music type (i.e. music −5, mood 3). It was also considered, that Mood might have different influence on different age groups, thus both youngest and oldest group of subjects were analysed separately. Values of attribute mood were discretized into 2 value groups representing negative (relaxed mood representing value range of {−5−0}) and positive (energetic mood representing value range of {0−5}) sides of mood value vector. In the next step data set has been divided into 2 separated ones containing data of younger and older groups of subjects. Then, both groups of subjects were filtered for attribute selection using WEKA Exhaustive Search and the Cfs Subset Evaluator [5]. For group of young adults reduction of attributes resulted in selection of Mood, Latency, Amplitude and Pupil Size and group of older adults had the same attribute selection. Dataset of younger adults contained 229 instances and dataset of older adults contained 145. Statistics are presented in Tables 3 and 4.

Table 3. Means of attribute Mood in age group of younger adults {<30}.

Mood groups	Latency	Amplitude	Avg velocity	Pupil size
−5–0	251.5 ms ± 50.4 (SD)	5.0 deg ± 0.6 (SD)	95.3 deg/sec ± 16.8 (SD)	7.3 mm ± 1.3 (SD)
0–5	248.9 ms ± 64.8 (SD)	5.2 deg ± 0.8 (SD)	98.3 deg/sec ± 18.0 (SD)	7.0 mm ± 1.2 (SD)
df	226	226	226	
Standard error of difference	7.689	0.094	2.306	0.166
P	0.7356	0.0338	0.1946	0.0715

Table 4. Means of attribute Mood in age group of older adults {>60}.

Mood groups	Latency	Amplitude	Avg velocity	Pupil size
−5–0	316.4 ms ± 84 (SD)	5.2 deg ± 0.6 (SD)	100.4 deg/sec ± 12.8 (SD)	6.2 mm ± 1.4 (SD)
0–5	326.5 ms ± 77 (SD)	4.8 deg ± 0.6 (SD)	93.4052 ± 16.3 (SD)	5.6 mm ± 1.3 (SD)
df	142	142	142	
Standard error of difference	13.429	0.100	2.442	0.225
P	0.4532	0.0001	0.0048	0.0086

According to the results attribute Latency has been found statistically insignificant. Also significant difference was observed in the Amplitude among younger group and in the Amplitude, the Average Velocity and the Pupil Size among older group. Interestingly, in group of older adults pupils were 0.5 mm larger for sessions with relaxed mood, while in group of younger adults pupil size seems to be static and not fragile for mood changes. 0.6 mm is equal to the change by 13% in pupil's size scale defined by possible range of 2–8 mm. According to this, saccades performed by subjects over age 60, with respectively different average pupil size, could be assigned to the set of results for one of side of the mood vector. It seems that in the elderly, pupil size and latency plays more important role in distribution of saccade parameters depending on mood, perhaps because they present with abnormalities connected to age or disease. Obtained results proves that Latency and Pupil Size, along with Amplitude and

Average Velocity could be found as differentiators between saccades made by people in different age and moods. Thus theoretically, age and mood of particular subject could be predicted from parameters of saccades. For this purpose, accuracy of model presented in the "Results" section has been tested against collected data with different types of classifiers: Random Forest (RF), Multilayer Perceptron (MLP), Naive Bayes (NB) and Support Vector Machine (SVM), in same WEKA's procedure of cross-validation of 10 folds [5]. Algorithm used in WEKA divides dataset into 10 folds and perform tests and train procedure for each fold with the remaining 9, giving in the result the average of 10 evaluations [5]. From classifiers enumerated above, two classifiers - RF and MLP - showed highest and almost equal accuracy.

For classification of Age groups, best results has been obtained for dataset containing all attributes collected during the experiment and with attribute Age discretized into 2 bins, matching old and young adults groups and also with attribute Mood discretized into 10 bins, matching granular values of the mood vector. For such dataset, both MLP and RF showed the same accuracy of prediction of 91.4% (correctly classified instances). In case of the Mood groups, best result also has been obtained for dataset containing all attributes, with attribute Age discretized into 44 bins (23 was the youngest and 67 was the oldest subject) and also with attribute Mood discretized into 2 bins, representing 2 sides of the mood vector. For MLP, the percentage of correctly classified instances was 96.6%, while RF performed slightly better with 97% correctly classified instances.

5   Discussion and Conclusions

The purpose of this experiment was to show that depending on the age, emotions affect saccade parameters (thus, the eye movement) in different ways and that these parameters can be used to classify subject to one of the defined group of age and mood. We also wanted to describe main differences between groups in our results. When comparing means of people under 30 with people over 60, significant difference can be noticed in ranges of values for attributes Latency and Average Pupil Size (Table 2). According to the results, mean latency of older people is ~23% (~77 ms) slower than among youngest group. The results are also different in pupil size. According to the results, average pupil size among older people is ~17% (1 mm) smaller, comparing to age group below 30. Both attributes has been found statistically significant. This findings and size of the difference seems to be consistent with results obtained by researchers mentioned earlier in this article, stating that only those two parameters are sensible for age related changes [11,12,17].

According to the results, latency seems to be the most sensitive to the age of all saccade parameters and variability of latencies is an interesting phenomenon of which mechanics has been described at least by two theoretical models: LATER (Linear Approach To Threshold With Ergodic Rate) [7] and SAT (Speed–Accuracy Tradeoff) [1,14,16].

Mechanics presented by both models are based on theory of accumulation of sensory evidences. So, according to our results one can hypothesize, that older people may need more time to accumulate enough evidences to commit decision with appropriate accuracy. This response time is defined by threshold of sensory input which might be different in older people, due to greater length of sampling, computation or communication process between brain areas. Those differences may have source in changes in brain due to age or neurodegeneration, but it may also come from the developing disease. Latency parameter of saccades seems to be most sensitive to these changes and shows how different brain processing is among group of older people.

When comparing influence of music expressed in attribute mood, we can see differences in attributes significance between both age groups. Among people between 20–30 years old (Table 3) actually only attribute Amplitude can be considered as marginal difference between relaxation (lower values) and arousal (higher values). Among older people, attributes significance looks slightly different. We can see that attributes Amplitude, Average Velocity and Pupil Size can be considered statistically significant (Table 4). Means of pupil size in group of people over 60 are different between moods by ~13%. This difference also occurs in group of people below 30, however is not significant. This result partially follows findings made by Gingras, Marin, Puig-Waldmüller and Fitch in their research on young people pupillary responses to the music [2]. However in case of this experiment, music was played before, not during measurement sessions to avoid interaction with ANS. So strong, post-music listening pupillary response found only in older people seems to be an interesting output for further investigations.

In case of Amplitude and Average Velocity, in group of people over 60 we can see much stronger significance, when comparing to group of young adults. Interestingly, values of both parameters are lower for energetic mood. This phenomenon is not following the widespread assumption, stating that arousal increases saccade amplitude and velocity. However there are several factors that could decrease saccade performance in older adults despite the state of arousal. It could be level of fatigue, decreased task difficulty due to retrain or decreased motivation, reward or effort, as sessions with energetic music was performed always at the end of the experiment [13].

The experiment showed that same task repeated in different emotional conditions by people in different age brings different outputs. Our results showed that in case of saccades, it appears to be possible to track down differences and classify subjects to defined ranges of age or mood with satisfying prediction ratio. This proves the sense of experiments involving the classification of the subject based on oculometric measurements.

Acknowledgements. This work was partly supported by projects Dec-2011/03/B/ST6/03816 from the Polish National Science Centre.

References

1. Abrams, R.A., Meyer, D.E., Kornblum, S.: Speed and accuracy of saccadic eye movements: characteristics of impulse variability in the oculomotor system. J. Exp. Psychol. Hum. Percept. Perform. **15**(3), 529–543 (1989). https://doi.org/10.1037/0096-1523.15.3.529
2. Gingras, B., Marin, M.M., Puig-Waldmüller, E., Fitch, W.T.: The eye is listening: music-induced arousal and individual differences predict pupillary responses. Front. Hum. Neurosci. **9**, 619 (2015). https://doi.org/10.3389/fnhum.2015.00619
3. Winn, B., Whitaker, D., Elliott, D.B., Phillips, N.J.: Factors affecting light-adapted pupil size in normal human subjects. Invest. Ophthalmol. Vis. Sci. **35**(3), 1132–1137 (1994)
4. Purves, D., Augustine, G.J., Fitzpatrick, D., et al. (eds.): Types of Eye Movements and Their Functions - Neuroscience, 2nd edn. Sinauer Associates Inc., Sunderland (2001). https://www.ncbi.nlm.nih.gov/books/NBK10991/
5. Hall, M., Frank, E., Holmes, G., Pfahringer, B., Reutemann, P., Witten, I.H.: The WEKA data mining software: an update. SIGKDD Explor. **11**(1), 10–18 (2009)
6. Huette, S., Winter, B., Matlock, T., Ardell, D.H., Spivey, M.: Eyemovements during listening reveal spontaneous grammatical processing. Front. Psychol. **5**, (2014). https://doi.org/10.3389/fpsyg.2014.00410
7. Imran, N.: Later models of neural decision behavior in choice tasks. Front. Integr. Neurosci. **8**, 67 (2014). https://doi.org/10.3389/fnint.2014.00067
8. Ellis, R.J., Thayer, J.F.: Music and autonomic nervous system (dys) function. Music Percept. **27**(4), 317–326 (2010). https://doi.org/10.1525/mp.2010.27.4.317
9. Pratt, J., Abrams, R.A., Chasteen, A.L.: Initiation and inhibition of saccadic eye movements in younger and older adults an analysis of the gap effect. J. Gerontol. Psychol. Sci. **52B**(2), 103–107 (1997)
10. Jacobson, D.M.: Relationship between age and pupil size (2002). http://content.lib.utah.edu/cdm/ref/collection/EHSL-Moran-Neuro-opth/id/105
11. Kaiser, P.K.: The Joy of Visual Perception. York University, Toronto (1996). http://www.yorku.ca/eye/thejoy.htm
12. Karageorghis, C.I., Priest, D.L.: Music in the exercise domain: a review and synthesis (Part I). Int. Rev. Sport Exerc. Psychol. **5**(1), 44–66 (2012). https://doi.org/10.1080/1750984X.2011.631026
13. Di Stasi, L.L., Catena, A., Cañas, J.J., Macknik, S.L., Martinez-Conde, S.: Saccadic velocity as an arousal index in naturalistic tasks. Neurosci. Biobehav. Rev. **37**(5), 968–975 (2013). https://doi.org/10.1016/j.neubiorev.2013.03.011
14. Marcus, N., Kenneth, H.: An adaptive algorithm for fixation, saccade, and glissade detection in eye tracking data. Behav. Res. Methods **42**(1), 188–204 (2010). https://doi.org/10.3758/BRM.42.1.188
15. Shafiq-Antonacci, R., Maruff, P., Whyte, S., Tyler, P., Dudgeon, P., Currie, J.: The effects of age and mood on saccadic function in older individuals. J. Gerontol. Psychol. Sci. **54**(6), 361–368 (1999)
16. Richard, H.: The speed-accuracy tradeoff: history, physiology, methodology, and behavior. Front. Neurosci. **8**, 150 (2014). https://doi.org/10.3389/fnins.2014.00150
17. Robinson, D.A.: Control of eye movements. Handbook of physiology, Section I: The nervous system. American Physiological Society: Bethesda, MD (2011). https://doi.org/10.1002/cphy.cp010228
18. Schafer T. Fachner, J.A.: Listening to music reduces eye movements. Percept. Psychophys. **77** (2015). https://doi.org/10.3758/s13414-014-0777-1

Algorithms and Optimization

Using Copula and Quantiles Evolution in Prediction of Multidimensional Distributions for Better Query Selectivity Estimation

Dariusz Rafal Augustyn[(✉)]

Institute of Informatics, Silesian University of Technology,
Akademicka 16, Gliwice, Poland
draugustyn@polsl.pl

Abstract. In query optimization theory a selectivity parameter is used by cost query optimizer for early estimating the size of data that satisfies a query condition. It requires some representation of distribution of attribute values. There are many approximate representations of m–d distribution where the copula-based is new one. This approach gives a possibility to take into account the fact of a varying m–d distribution by predicting both a copula and 1–d marginal distributions. In this paper we propose the method of forecasting trajectories of either copula parameters and marginals' quantiles using time series prediction models. This method is mainly designated for predicting outdated distribution representation what may improve accuracy of selectivity estimation based on such representation. It also may be used for predicting a varying query workload to forecast important regions of data domain. Having detected such regions we may improve there the resolution of distribution representation.

Keywords: Query optimization · Selectivity estimation · Multidimensional distribution · Quantiles evolution · Copula · Time series prediction

1   Introduction

Selectivity estimating belongs to query optimization process. A query selectivity value is needed by a database module – a cost query optimizer – for selecting the best query execution plan. Having a fraction of table rows that satisfy query conditions, the optimizer selects a method of data access to a table (full-scan, index-based, ...) or an order of tables joins. Selectivity for a single-table query is a number of rows that satisfy a query condition divided by number of all table rows. Selectivity is also an estimator of probability of drawing a row which satisfies the query condition.

© Springer Nature Switzerland AG 2020
A. Gruca et al. (Eds.): ICMMI 2019, AISC 1061, pp. 209–220, 2020.
https://doi.org/10.1007/978-3-030-31964-9_20

A selectivity value for a range query Q is defined as follows:

$$
\begin{aligned}
&sel(Q(a_1 < x_1 < b_1 \wedge \cdots \wedge a_D < x_D < b_D)) \\
&= \int_{a_1}^{b_1} \cdots \int_{a_D}^{b_D} f_{X_1 \ldots X_D}(x_1, \ldots, x_D) dx_1 \ldots dx_D
\end{aligned}
\tag{1}
$$

where $f_{X_1 \ldots X_D}(x_1, \ldots, x_D)$ is a multivariate probability density function (PDF) for a joint distribution of continuous-domain attributes $x_1, \ldots x_D$ and a_d, b_d are boundaries of ranges for each x_d for $d = 1 \ldots D$.

We can see in Eq. (1) that selectivity calculations requires some estimator of the multivariate PDF. Naïve approaches could be based on a multidimensional equi-width histogram. The multidimensional domain of such histogram consists of m-d equi-width intervals, i.e. hyper-rectangles with an equal length of edges in each dimension. It is rather obvious that such approach is very space-consuming for high dimensions. This is known as so-called the curse of dimensionality. More space—efficient approach is needed. Thus database management system (DBMS) stores and uses only some representation of a marginal distribution for values of attribute x_d (i.e. an estimator of an univariate PDF $f_{X_d}(x_d)$). To obtain a selectivity of query Q with condition based on many attributes, DBMS calculates partial selectivities for each simple predicates Q_d based on a single attribute, and then multiply calculated values for obtaining the result selectivity value:

$$
\begin{aligned}
&sel(Q(a_1 < x_1 < b_1 \wedge \cdots \wedge a_D < x_D < b_D)) \\
&\approx \prod_{d=1}^{D} sel(Q_d(a_d < x_d < b_d)) = \prod_{d=1}^{D} \int_{a_d}^{b_d} f_{X_d}(x_d) dx_d.
\end{aligned}
\tag{2}
$$

Such approach is based on Attribute Values Independence - AVI rule [7] which is not satisfied in correlated data.

Generally, DBMS optimizers rather utilize AVI method which uses partial selectivites obtained using equi-width or equi-height 1-d histograms that individually describe 1-d marginal distributions. But there are known some more accurate approaches that based on a joint distribution and its multidimensional representation (e.g. ones based on Cosine Transform [8], Wavelets [4], Bayesian Networks [5], m-d Copula [3]). All of them are based implicitly on concept of a lossy compressed representation of m-d distribution.

m-d Copula approach may be useful as a minimal space-consuming method but still multidimensional one in contrast to AVI method. It has a better accuracy than AVI method for correlated data [3], rather not better accuracy than the others (because they may use more metadata for describing m-d distribution, depending on established degree of compression) but it may be the least space-consuming among the others. It uses small size of metadata i.e. only 1-d histograms for marginal distributions and few parameters that describe a multidimensional joint dependency.

Another aspect of selectivity estimation is connected with a problem of selectivity estimation based on a partially outdated representation of distribution because of changing database. When the cost of updating the distribution representation is large we can use some approaches based on predicting a future representation using historical existing ones. Such approach focused on a 1-d representation was presented in [2].

The main goal of this work is about proposing a space-efficient method of distribution representation for changing m-d data. The method allows to predict a representation in near future. Because of necessity of holding a few m-d representations made in a few previous moments of time, the method should be possibly the least space-consuming. It may be achieved by applying either the m-d Copula-based method for selectivity calculations or prediction models only for few Copula's parameters and few interval boundaries of histograms describing only 1-d marginal distributions.

In this paper we propose the method which allows to calculate a copula-based selectivity estimator for changing multidimensional data. The main contributions of the method described in the paper are:

- the concept of forecasting a m-d distribution for more accurate selectivity estimation by applying the method of predicting evolution of either copula parameters (for extrapolation a joint m-d data dependency) or quantiles (for extrapolation 1-d marginal distributions),
- the concept of usage of space-saving representation of statistics for predicting a low resolution representation of distribution and then improving its final resolution by interpolation.

2   Theoretical Background

2.1   Copula in Selectivity Estimation for m-d Range Query

A multidimensional copula C [6] is a joint cumulative distribution function (CDF) of a D-dimensional random vector $[U_1, \ldots, U_D]$ on the unit cube $[0,1]^D$ with uniform marginals:

$$C(u_1, \ldots, u_D) = P(U_1 \leq u_1, \ldots, U_D \leq u_D). \tag{3}$$

A copula density is defined as as a high order partial derivate of C:

$$c(u_1, \ldots, u_D) = \frac{\partial^D C(u_1, \ldots, u_D)}{\partial u_1 \ldots \partial u_D}. \tag{4}$$

Any m-d CDF $F_{X_1, \ldots, X_D}(x_1, \ldots, x_D)$ can by expressed using m-d C and marginal CDFs $F_{X_d}(x_d)$ for $d = 1 \ldots D$:

$$F_{X_1, \ldots, X_D}(x_1, \ldots, x_D) = C\left(F_{X_1}(x_1), \ldots, F_{X_D}(x_D)\right). \tag{5}$$

Using Eq. (5), we may obtain copula from multivariate CDF and univariate inverted CDFs as follows:

$$C(u_1, \ldots, u_D) = F_{X_1 \ldots X_D}(F_{X_1}^{-1}(u_1), \ldots, F_{X_D}^{-1}(u_D)) \tag{6}$$

where $u_d = F_{X_d}(x_d)$ for $d = 1 \ldots D$.

Using $u_d = F_{X_d}(x_d), du_d = f_{X_d}dx_d, ua_d = F_{X_d}(a_d), ub_d = F_{X_d}(b_d)$ (see [3]) the selectivity may be calculated as a definite multi-integral of copula density as follows:

$$sel(Q(a_1 < x_1 < b_1 \wedge \cdots \wedge a_D < x_D < b_D))$$
$$= \int_{ua_1}^{ub_1} \cdots \int_{ua_D}^{ub_D} c(u_1, \ldots, u_D)du_1 \ldots du_D. \tag{7}$$

We can obtain a selectivity value without integrating of over D-dimensional hyper-rectangle defined by pairs (ua_d, ub_d) for $d = 1 \ldots D$ but using copula $C(u_1, \ldots, u_D)$ directly. Formulas for selectivity calculations for 2-d case and 3-d one are presented in [3].

In proposed method which based on approach presented in [3] where we used the well known bi-variate Clayton copula copula defined as follows:

$$C(u_1, u_2; \Theta) = \max[(u_1^{-\Theta} + u_2^{-\Theta} - 1)^{-\frac{1}{\Theta}}, 0]. \tag{8}$$

which has only one copula parameter $\Theta \in [-1, +\infty)\setminus\{0\}$.

The copula-based representation may be used for D-dimensional distribution especially by using nesting approach. To do this we can use D-variate nested copula with $D - 1$ parameters:

$$C(u_1, u_2, \ldots, u_D; \Theta_1, \Theta_2, \ldots, \Theta_{D-1})$$
$$= C_A(\ldots C_A(u_1, u_2; \Theta_1), u_3; \Theta_2) \ldots, u_D; \Theta_{D-1}) \tag{9}$$

where C_A is a bi-variate Clayton copula given by Eq. (8).

In further considerations that deal with applying the proposed copula-based method for changing distribution, a prediction of Θ parameters (from Eq. (9)) will be used to forecast an approximate representation of m-d distribution. Θ parameters will be used for modeling a future dependency of m-d data.

2.2 Quantile-Based Histogram

There are many kinds of histogram for representing 1-d probability density. One of them is called equi-depth (equi-height) histogram. It is a series of quantiles. Such histogram is a series of x_p values for p belonging to $(0, 1)$ and satisfying:

$$P_X(x \leq x_p) \geq p \wedge P_X(x \geq x_p) \geq 1 - p. \tag{10}$$

q-quantiles is a series of x_p for $p = 1/q, 2/q, \ldots, (q - 1)/q$. To take into account boundaries of x domain we will use additional very low order quantile and very high one, e.g. $p = 0.001$ and $p = 0.999$. Thus q-quantile-based histogram for continuous variable we define as a series of x_p values:

$$(x_p) : x_p = \int_{-\infty}^{x_p} f(x)dx = p \wedge p = 0.001, \frac{1}{q}, \frac{2}{q}, \ldots, \frac{q-1}{q}, 0.999 \tag{11}$$

for $q < 1000$. It will be called q-hist. Storing q-hist requires to hold $q+1$ values. In further considerations that deal with applying the proposed method for changing distribution, prediction of q-quantiles will be used to forecast an approximate representation of m-d distribution. q-quantiles will be used for modeling future 1-d marginal distributions.

2.3   Applied Prediction Methods

Prediction of distribution parameters may be based on many well-known approaches. In the proposed method we use simple discrete linear dynamic models like AR (autoregressive) and ARMA (autoregressive and moving average)[1]. Problems of nonstationarity, presence of trend, periodicity (that are related to the applicability of these models) are not discussed here but they could be taken into account in future considerations. Let's assume:

- $y(t), y(t-1), \ldots, y(t-n_a)$ - time series values,
- $e(t), e(t-1), \ldots, e(t-n_c)$ - white-noise disturbance

where t is an index that points a moment of time. For AR(n_a) model we have the following recurrent equation:

$$y(t) + a_1 y(t-1) + \ldots + a_{n_a} y(t-n_a) = e(t). \tag{12}$$

For ARMA(n_a, n_c) model we have the following recurrent equation:

$$y(t) + a_1 y(t-1) + \ldots + a_{n_a} y(t-n_a) = e(t) + c_1 e(t-1) + \ldots + c_{n_c} e(t-n_c). \tag{13}$$

Introducing the following polynomials:

$$A(z) = 1 + a_1 z^{-1} + \ldots + a_{n_a} z^{-n_a}, C(z) = 1 + c_1 z^{-1} + \ldots + c_{n_c} z^{-n_c}. \tag{14}$$

(where z is a delay operator) we can define AR model and ARMA one as follows:

$$\text{AR: } A(z)y(t) = e(t), \quad \text{ARMA: } A(z)y(t) = C(z)e(t). \tag{15}$$

Choosing the model is based on rating of model fitting where we use FPE coefficient (we choose the smallest value of Final Prediction Error). To find a new predicted value of y we need to use n_a previously stored coefficients of chosen model and n_a previous values of y. Above-mentioned models will be separately used for predicting values of each q-quantiles (in each dimension) and Θ values.

3   The Proposed Method of Predicting Representation of m-d Distribution and Selectivity Estimation

To present the proposed method we define:

- copula parameters $(\Theta_j)_{j=1}^{D-1}(t), (\Theta_j)_{j-1}^{D=1}(t-1), \ldots$ for moments of time: $t, t-1, \ldots$ and predicted one $(\Theta_j)_{j=1}^{D-1}(t+1)$ for the future moment $t+1$,

---

[1] Estimate AR and ARMA Models – Matlab and Simulink (2019) https://www.mathworks.com/help/ident/ug/estimating-ar-and-arma-models.html.

- low resolution histograms $q\text{-hist}_d(t)$, $q\text{-hist}_d(t-1)$, ... for each d-th dimension and for moments $t, t-1, ...$ and predicted one $q\text{-hist}_d(t+1)$,
- high resolution histograms $Q\text{-hist}_d(t)$ for the last moment of time t and interpolated one $Q\text{-hist}_d(t+1)$ for the future moment.

A high resolution histogram $Q\text{-hist}_d$ is based on $Q = kq$ number of quantiles (where k is small integer value). It is important to notice that $Q\text{-hist}_d$ contains $q\text{-hist}_d$. Thus $q\text{-hist}_d$ may be obtained from $Q\text{-hist}_d$ by taking only every k-th element from of $Q\text{-hist}_d$. For space-efficiency reason (according to $q < Q$) we store only low resolution histograms in previous moments of time.

Algorithm 01 obtains a new statistics (copula parameters and marginals estimators) directly from data for the given moment of time t.

Algorithm 01:
```
01: Obtain (Θ_j)_{j=1}^{D-1}(t) using the current data and store it
02: For each dimension d = 1...D:
03:    Obtain q-hist_d(t − 1) from Q-hist_d(t − 1) and store it
04:    Obtain Q-hist_d(t) using the current data and store it
```

Algorithm 02 predicts a new statistics for the future moment $t+1$.

Algorithm 02:
```
01: Using copula parameters (Θ_j)_{j=1}^{D-1}(t), (Θ_j)_{j=1}^{D-1}(t − 1), ...
  .    and applying the best prediction model (AR, ARMA, ...)
  .    predict (Θ_j)_{j=1}^{D-1}(t + 1) and store it
02: For each dimension d = 1...D:
03:    Using q-hist_d(t) (taken from Q-hist_d(t)) and q-hist_d(t − 1), ...
  .      and applying the best prediction model (AR, ARMA, ...)
  .      predict q-hist_d(t + 1) and store it
04: Using Q-hist_d(t) and predicted q-hist_d(t + 1)
  .      obtain Q-hist_d(t + 1) by interpolation (Eq. (16)) and store it
```

In the described approach we assume that we find a separate prediction model for each Θ (Algorithm 02 line 01). The order of the model for each Θ_j will be denoted by $n_a(\Theta_j)$ for $j = 1 ... D-1$. We also assume that we find a separate prediction model for any quantile belonging to $q\text{-hist}_d$ (Algorithm 02 line 03). The order of prediction model for each quantile will be denoted by $n_a(x_{d\,p})$ where p is defined in Eq. (11) and d is index of dimension.

Having high resolution $Q\text{-hist}_d(t)$ and low resolution $q\text{-hist}_d(t+1)$ we will calculate (Algorithm 02 line 04) missing values by distributing them among the $q\text{-hist}_d(t+1)$ values according the distribution from $Q\text{-hist}_d(t)$. It is shown in Fig. 1. where $v_1(t+1)$ and $v_6(t+1)$ are known (they are obtained during the prediction and included to $q\text{-hist}_d(t+1)$). The rest of values (not predicted) v_j where $j = 2, 3, 4, 5$ in $Q\text{-hist}_d(t+1)$ will be obtained by interpolation:

$$v_j(t+1) = v_1(t+1) + \frac{v_6(t+1) - v_1(t+1)}{v_6(t) - v_1(t)}(v_j(t) - v_1(t)). \qquad (16)$$

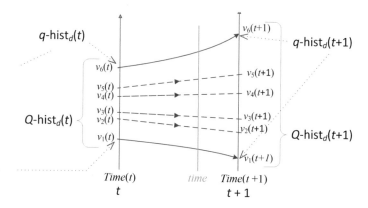

Fig. 1. Obtaining missing values $(v_2, \ldots, v_5)$ in predicted Q-hist$_d(t + 1)$ by interpolation.

Algorithm 03 allows to calculate a selectivity in any *time* between moments t and $t + 1$ (Fig. 1) by assuming a simple linear model of time dependency for any statistics:

$$stat(time) = \frac{stat(Time(t + 1)) - stat(Time(t))}{Time(t + 1) - Time(t)} time + stat(Time(t)). \quad (17)$$

where $Time(t)$ is continuous time value in a moment number t and $stats(Time(t))$ is any theta $(\Theta_j)_{j=1}^{D-1}(t)$ or any quantile-base histogram Q-hist$_d(t)$.

Algorithm 03:
```
inputs:
- time ∈ (Time(t), Time(t + 1)],
- range query boundaries ([a_d, b_d])^D_{d=1} (Eq. (1))
01: For each dimension d = 1...D:
02:    Using Q-hist_d(time) (Eq. (17))
.          obtain 1-d histogram - estimator of marginal PDF - f_d(x_d)
.          and then obtain 1-d estimator of marginal CDF - F_d(x_d)
03:    Obtain ua_d = F_d(a_d) and ub_d = F_d(b_d)
04: Using (Θ_j)^{D-1}_{j=1}(time) (Eq. (17)) and boundaries ([ua_d, ub_d])^D_{d=1}
.      obtain selectivity value (see Eq. (8)(9))and return it
```

4 Presenting the Method in Experimental Results

This section shows how the method works for some exemplary 2-dimensional varying distribution of pairs (x_1, x_2). Initially, we used reference Matlab's dataset - stockreturns[2], shifted to the right for achieving only nonnegative values (see Fig. 2a).

---

[2] Matlab Sample Data Sets (2016) https://www.mathworks.com/help/stats/_bq9uxn4.html.

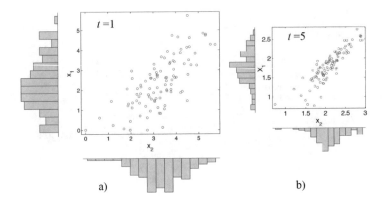

Fig. 2. Datasets of pairs (x_1, x_2): (a) initial dataset for $t = 1$, (b) the dataset for $t = 5$.

To simulate an evolution of dataset distribution we use some nontrivial transformation formula that consists of the linear component - $ax_i + b$, the nonlinear one - $\sqrt{(x_i)}$ and the correlation one - $x_i \cdot x_j$, which is applied to M data pairs for a sequence of moments of time $t = 1 \ldots T$:

$$x_{1,m}(t+1) = 0.5x_{1,m}(t) + 0.5\sqrt{x_{1,m}(t)} + 0.05x_{1,m}(t) \cdot x_{2,m}(t)/\overline{x_2}(t)$$
$$x_{2,m}(t+1) = 0.5x_{2,m}(t) + 0.5\sqrt{x_{2,m}(t)} + 0.05x_{1,m}(t) \cdot x_{2,m}(t)/\overline{x_1}(t). \tag{18}$$

where $\overline{x_d}(t) = \frac{1}{M}\sum_{m=1}^{M} x_{d,m}(t)$ is a mean value of variable x_d for $d = 1 \ldots D$ ($D = 2$) in the moment t.

Learning datasets were generated for $t = 1 \ldots T$ where $T = 5$. The dataset for $t = 5$ was shown in Fig. 2b. The dataset for $t = 6$ will be used as a testing dataset.

Because of only 2-dimensional data we will consider only one Clayton copula parameter $\Theta(t)$ - for all dataset. By executing Algorithm 01 line 01 for moments: $t = 1 \ldots 5$ we obtain thetas learning sequence presented in Fig. 3a, b.

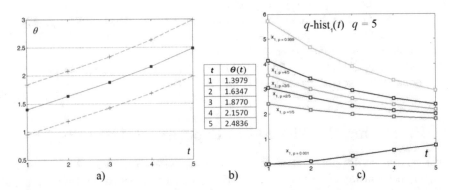

t	$\Theta(t)$
1	1.3979
2	1.6347
3	1.8770
4	2.1570
5	2.4836

Fig. 3. Statistics for $t = 1 \ldots 5$ gathered from learning datasets: (a) $\Theta(t)$ and its' confidence intervals for 0.95 confidence level, (b) list of Θ values, (c) evolution of quantiles for variable x_1 – varying of q-hist$_1(t)$ for $q = 5$.

Having five previous theta values (Fig. 3b) we may obtain (Algorithm 02 line 01) linear prediction models shown in Table 1.

Table 1. Rating of models for prediction of Θ based on values: $\Theta(t = 1), \ldots, \Theta(t = 5)$.

Model	Polynomial	n_a	Loss function	FPE
AR(1)	$A(z) = 1 - 1.153z^{-1}$	1	0.000129655	0.000181517
AR(2)	$A(z) = 1 - 1.032z^{-1} - 0.1356z^{-2}$	2	2.18573e-006	3.93431e-006
ARMA(1,1)	$A(z) = 1 - 1.15z^{-1}$ $C(z) = 1 + 0.7537z^{-1}$	1	4.88741e-006	1.14039e-005

Taking into account the smallest FPE value from Table 1 we can choose AR(2) as the best model and find the optimal model:

$$\Theta(t + 1) = 1.032\Theta(t) - 0.1356\Theta(t - 1) \qquad (19)$$

which allows to obtain predicted value of $\Theta(t = 6) = 2.8556$ by using theta values from Fig. 3b. A true value of $\Theta(t = 6)$ (obtained from test dataset by applying Eq. (18) for $t = 6$) is equal 2.8660. Thus, the relative prediction error of $\Theta(t = 6)$ equals only about 0.4% $\approx (2.8660 - 2.8556)/2.8660 \cdot 100\%$.

For modeling marginal distributions we used $q = 5$ as an order of quantiles for low resolution q-histograms and $Q = 10$ for high resolution Q-histograms.

Executing Algorithm 01 line 03 for $t = 1 \ldots 5$ allows to obtain quantiles for x_1 and x_2, i.e. obtain q-hist$_1(t)$ and q-hist$_2(t)$. The evolution of quantiles for x_1 i.e. q-hist$_1(t)$ – was shown in Fig. 3c.

By executing Algorithm 02 line 03 we choose the best-fitting AR or ARMA model using quantile's values from $t = 1 \ldots 5$ and predict each quantile value in $t = 6$. This allows to obtain predicted q-hist$_1(t = 6)$ and q-hist$_2(t = 6)$. The result predicted q-hist$_1(t = 6)$ is shown in Fig. 4a.

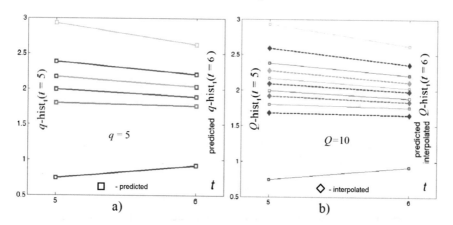

Fig. 4. Obtaining predicting Q-hist for variable x_1: (a) prediction of low resolution q-hist$_1(t = 6)$ from previous q-hist$_1(t)$ for $t = 1 \ldots 5$), (b) interpolation (dashed lines) – obtaining missing values for predicted and interpolated high resolution Q-hist$_1(t = 6)$ by using q-hist$_1(t = 6)$ and Q-hist$_1(t = 5)$.

To obtain the high resolution histogram Q-hist$_d(t = 6)$ we used interpolation (see Fig. 1) using quantiles from Q-hist$_d(t = 6)$. Assuming $k = 2$ this allows to double a number of quantiles in Q-hist$_d(t = 6)$ comparing to the predicted q-hist$_d(t = 6)$. The resulting predicted and interpolated Q-hist$_1(t = 6)$ was shown in Fig. 4b.

Quantiles in predicted histograms Q-hist$_d(t = 6)$ for $d = 1 \ldots 2$ are used for obtaining estimator of PDF – f_d and then marginal CDF – F_d for each dimension. CDFs are directly used in selectivity calculation in Algorithm 03 line 03. The functions $f_1(x_1)$ and $F_1(x_1)$ obtained using predicted Q-hist$_1(t = 6)$ are shown in Fig. 5.

To verify the proposed method of query selectivity estimation we will use synthetic sample set of 2-d range queries $Q_j(a_{1j} < x_1 < b_{1j} \wedge a_{2j} < x_2 < b_{2j})$ for $j = 1 \ldots J_{max}$ and $J_{max} = 10000$. Values of query boundaries was generated independently for each dimension and with accordance to the truncated uniform distribution shown in Fig. 6 and scaled to the min and max values of x_1 and x_2.

Here we remind from [3] the approach to measure errors for given selectivity estimation method and for a series of queries Q_j:

$$Err_{meth}(Q_j) = \begin{cases} \frac{|sel_{meth}(Q_j) - sel_{true}(Q_j)|}{sel_{true}(Q_j)} & \wedge \quad sel_{true}(Q_j) \neq 0 \\ |sel_{meth}(Q_j)| & \wedge \quad sel_{true}(Q_j) = 0 \end{cases} \quad (20)$$

$$MeanErr_{meth} = \frac{\sum_{j=1}^{J_{max}} Err_{meth}(Q_j)}{J_{max}}.$$

Err_{meth} is a relative error if true selectivity does not equal 0 or an absolute error otherwise.

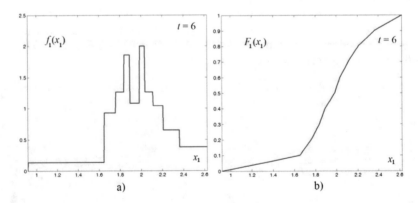

a)   b)

Fig. 5. (a) Predicted estimator of PDF for x_1 constructed using predicted and interpolated Q-hist$_1(t = 6)$, (b) predicted estimator of CDF for x_1 obtained from corresponding PDF.

The results show that the proposed Copula-based method with prediction is more accurate in about $10\% \approx (0.34 - 0.26)/0.34 \cdot 100\%$ relative to the Copula-based one without prediction and about **50\%** $= (0.52 - 0.26)/0.52 \cdot 100\%$ relative to the reference AVI-based one.

Table 2. Selectivity Mean Error measured for $t = 6$.

Method	MeanErr [%]
AVI-based *meth.* using only marginals for $t = 5$	0.52
Copula-based *meth.* using outdated statistics for $t = 5$ (**no prediction**)	0.34
Copula-based meth. with predicting only quantiles (**no prediction of theta**; theta taken for $t = 5$)	0.29
Copula-based *meth.* with predicting either quantiles or theta	0.26

After the learning phase based on the data for $t = 1 \dots 5$, we stated that all orders of prediction model either for quantiles or theta are equal 2 (i.e. $n_a(\Theta) = 2$ and $n_a(x_{d\,p}) = 2$ for all $p = 0.001, 1/5, 2/5, 3/5, 4/5, 0.999$ and all $d = 1, 2$). This means that for future steps of prediction we need to store $n_a(\Theta) + D \cdot (q + 1) \cdot n_a(x_{d\,p}) = 2 + 2 \cdot (5 + 1) \cdot 2 = 26$ values of model coefficients (The prediction models have the same orders but probably different coefficients). To calculate a selectivity we need current $Q\text{-hist}_1(t)$ and $Q\text{-hist}_2(t)$ and two additional predicted and interpolated $Q\text{-hist}_1(t+1)$ and $Q\text{-hist}_2(t+1)$. Storing those two last ones requires to hold $2 \cdot (Q+1) = 2(2q+1) = 2 \cdot (2 \cdot 5 + 1) = 22$ values. We also need current $\Theta(t)$ and predicted $\Theta(t+1)$ so we have to store two more values. Thus, the proposed method can be used for selectivity calculation or predicting new statistics for next moments of time in expense of additional size that equals $26 + 22 + 2 = 50$ values. This can be compared to the size of representation of statistics needed for the reference AVI-based method which requires metadata with size equals $22 = D \cdot sizeof(Q\text{-hist}_d) = 2 \cdot 11 = 22$. It is an increase of about $227\% \approx 50/22 \cdot 100\%$ of occupied space.

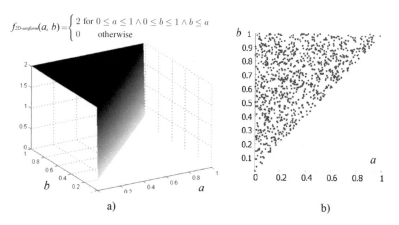

Fig. 6. Defining a testing set of range queries [1,3]: (a) 2-d truncated uniform PDF defining the distribution of query boundary in each dimension, (b) a sample data set of pairs (a, b) in selected dimension.

5   Conclusions

In the paper we considered a problem of selectivity estimation of range queries that operate on m-d data while the data are changed but outdated representation of distribution cannot be updated frequently enough. In such terms a method based on prediction of distribution of m-d data may be applied. We proposed the method which uses a predicted space—efficient representations based on m-d copulas. The method builds models for predicting time trajectories of either copula parameters or quantiles that describe marginals.

It is worth noticing that the described approach can be used not only for predicting distribution of the data but also for predicting query conditions, i.e. predicting m-d representation of boundaries of query ranges. This may be useful for solutions like [1] where we create such representations of data distribution that are aware of varying query workload. By using the proposed method we may predict regions of the user's interests. We may select such regions of data domain that need an improvement of resolution of distribution representation.

Applying the method gives us a promising results in an accuracy of selectivity estimation based on predicted statistics despite the increase in space occupancy.

Acknowledgements. This work was supported by the Statutory Research funds of Institute of Informatics, Silesian University of Technology (grant No BK/204/RAU2/2019).

References

1. Augustyn, D.R.: Query-condition-aware V-optimal histogram in range query selectivity estimation. Bull. Pol. Acad. Sci. Tech. Sci. **62**(2), 287–303 (2014)
2. Augustyn, D.R.: Using the model of continuous dynamical system with viscous resistance forces for improving distribution prediction based on evolution of quantiles. In: Proceedings of BDAS 2014, pp. 1–9 (2014)
3. Augustyn, D.R.: Copula-based module for selectivity estimation of multidimensional range queries. Man. Mach. Interact. **5**, 569–580 (2018)
4. Chakrabarti, K., Garofalakis, M., Rastogi, R., Shim, K.: Approximate query processing using wavelets. Int. J. Very Large Data Bases **10**(2–3), 199–223 (2001)
5. Getoor, L., Taskar, B., Koller, D.: Selectivity estimation using probabilistic models. In: ACM SIGMOD Record, vol. 30, pp. 461–472. ACM (2001)
6. Joy, H.: Dependence Modeling with Copulas. CRC Press, Bosa Roca (2015)
7. Poosala, V., Ioannidis, Y.E.: Selectivity estimation without the attribute value independence assumption. VLDB J., 486–495 (1997)
8. Yan, F., Hou, W.C., Jiang, Z., Luo, C., Zhu, Q.: Selectivity estimation of range queries based on data density approximation via cosine series. Data Knowl. Eng. **63**(3), 855–878 (2007)

Audio-Visual TV Broadcast Signal Segmentation

Josef Chaloupka[(✉)]

The Institute of Information Technology and Electronics Technical University
of Liberec, Liberec, Czech Republic
`josef.chaloupka@tul.cz`

Abstract. Research in the field of audio-visual broadcast programs transcription and indexing has been solved for more than 20 years. Great progress has been made mainly in the area of broadcast transcription from audio signal. In the last 10 years, this research has become more intense, mainly due to the use of deep or convolutional neural networks and because of large IT companies (Google, Microsoft, IBM, Amazon) that can rely on a large number of custom embedded multimedia databases. Very important part of system for audio-visual broadcast signal transcription is subsystem for signal segmentation. Signal segmentation is usually solved separately for audio and visual signal. In this paper, a methodology for audio-visual broadcast signal segmentation is presented and described. The result from audio signal segmentation is used for improving of visual signal segmentation.

Keywords: Audio-visual broadcast transcription ·
Audio-visual signal segmentation · LVCSR

1 Introduction

Now in 2019, systems of automatic processing, recognition and synthesis of speech acoustic signals are used in many areas [3]—computer dictation, voice control of computers, dialog systems, automatic transcription of radio and television programs, speech synthesis at railway stations and information systems, etc. Significant advances have been made in speech recognition in flexible languages, i.e. languages where word terminations are more or less varied depending on declination, verbs, adjectives, and adverbs. These languages include the Slavic languages, spoken by about 293 million people in the world. For flexible languages, large vocabularies must be used for speech recognition, rather than inflexible languages (e.g. English). Also, the relevant language models are considerably larger, resulting in greater computational demands.

In our Laboratory of Computer Speech Processing at the Technical University of Liberec, we have developed our own Large Vocabulary Continuous Speech Recognition (LVCSR) systems that work in real time with vocabularies containing more than 500,000 words. These systems were originally designed for

© Springer Nature Switzerland AG 2020
A. Gruca et al. (Eds.): ICMMI 2019, AISC 1061, pp. 221–228, 2020.
https://doi.org/10.1007/978-3-030-31964-9_21

Fig. 1. A prototype of our audio-visual broadcast transcription system.

the Czech language [14]. In the last 10 years, however, they have been adapted to practically all Slavic languages: Slovak [13], Polish [11], Russian, Belarusian, Ukrainian, Serbian, Croatian, Slovenian, Bulgarian and Macedonian [12]. For each Slavic language, a vocabulary, a language model, and a hybrid acoustic model have been specially designed and created. The actual acoustic model is based on Hidden Markov Models (HMM) and Deep Neural Networks (DNN) [2].

Our speech recognizer is already practically applied in a system for voice dictation to the computer and in a system for automatic transcription of television and radio programs. The practically achievable and experimentally verified recognition score for the voice dictation system is over 98%. For the system for automatic transcription of broadcast programs, this recognition score is around 86%. Here the recognition score is lower, as audio recordings often contain heavy noise or background music, or people use colloquial speech. In this transcription system, a broad field of scientific and research activity is still open to improve it.

The output of the transcription system for broadcast programs is an audio signal transcribed in text form together with time stamps where the individual recognizable words in the audio signal were found. Part of the system includes modules for detecting of speech/non-speech parts and a module for detecting the change of speaker. All of the output information is stored in output script and subsequently indexed in databases. Our transcription system has been fully

Visual signal

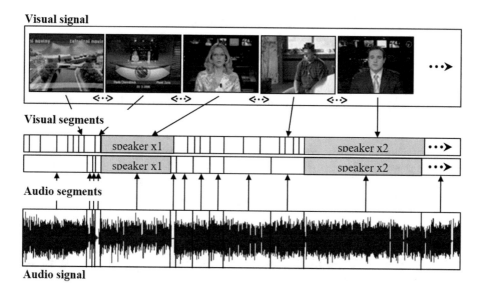

Fig. 2. Audio and visual segments in TV news

deployed and tested in the project of our Ministry of Culture, where more than 100,000 h of audio recordings from Czech Radio historical archive were transcribed and indexed [10].

In the last 10 years, the systems for broadcast transcription are being expanded with the modules for visual signal transcription if video recordings are available. These systems are being developed mainly by large IT companies.

Research in area of visual broadcast transcription is primarily focused on recognizing of interesting information in video recordings. If a person is speaking then his face is also captured. This is solved in the role of audio-visual speech recognition, where the visual part can improve the resulting recognition rate, especially in the noisy conditions. In addition to audio-visual speech recognition, human faces can be detected and identified in the visual part of the video recordings. The age, gender, and emotions of the person can be determined from the detected face. A very interesting and desired area today is the detection and identification of advertising logos and a number of other objects, which can ultimately serve to describe the original multimedia (video) data.

So, based on our experience in research area of speech and image processing and recognition, we also started to develop fully-featured audio-visual broadcast transcription system, see Fig. 1. or www: http://kvap.tul.cz/avminer.html At present, the audio-visual transcription system consists of several modules for: LVCSR, audio signal segmentation, audio speaker identification, visual signal segmentation, face detection and identification, Optical Character Recognition (OCR) of video frames and TV Logo recognition.

2   Audio-Visual Signal Segmentation

Very important part of the audio-visual broadcast transcription system is module for visual signal segmentation. A processing of each video frame would be rather computationally demanding therefore only Key-frames from visual segments are processed. The goal is to find positions (boundaries) of single visual segments. The boundaries of visual and audio segments are the same in some special cases mainly if the video is composed from various different parts. For example, these special cases occur quite often in television news, see Fig. 2. The main idea of the work described in this paper was to use the results of audio segmentation to train a new audio-visual model for visual signal segmentation. The assumption was that the resulting visual segmentation could be improved.

2.1   Audio Signal Segmentation

A Speaker Change Point (SCP) detection together with Speech Activity Detection (SAD) is used for audio signal segmentation very often. SCP is more challenging task than SAD therefore the research focuses mainly on the design of sufficiently reliable system for SCP. I-vectors [9], Deep Neural Networks (DNNs) [16] or CNNs [4] are implemented in state-of-the-art systems at present and the output signal is the most often parameterized into Mel-Frequency Cepstral Coefficients (MFCCs) [17].

We have designed and created our own SCP system for real-time broadcast monitoring based on CNN where an input is pre-processed by SAD based on DNN and Weighted Finite State Transducer [8]. The CNN was trained on artificial dataset where 20000 audio recordings (approx 30 h) with homogeneous information (only one speaker speaks) were connected. So, a position of speaker change point was known. The CNN was composed of two convolutional and two fully connected layers. An input audio signal was parameterized into 13 MFCCs + delta and delta-delta features. The resulting feature vector has size of 39 audio features. The input to CNN was 2 second window from input signal where 1 second consists of 100 parameterized frames. The input to CNN therefore consisted of 201 (200 + 1 frame) feature maps (39×1 in size).The first convolutional layer had 105 feature maps (39×1 in size) followed by 3:1 max-pooling layer. The second convolutional layer had 157 feature maps (13×1 in size) and the fully connected layers consisted from 64 neurons per layer, see Table 1. The CNN was trained in 15 epochs.

Our SPC system based on CNN was compared by result from LIUM speaker diarization toolkit [15]. Well known F-measure value was used for measure of quality of SCP detection. Both systems were tested on Czech subset from publicly available COST278 broadcast news database [19]. It was 399 speaker change points in 90 min of subset dataset. The F-measure from our system was 85,6%. Reference LIUM system had F-measure 84,6%.

Table 1. CNN architecture for audio signal segmentation

Layer	Architecture	Transfer function
Convolution	105 kernels 39 × 1	ReLU
Convolution	157 kernels 13 × 1	ReLU
Fully connected	64 neurons	ReLU
Output	2 neurons	SoftMax

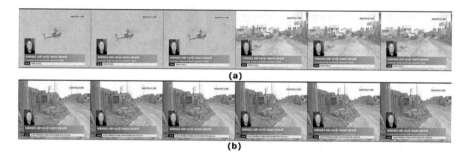

Fig. 3. 6-video frame image (a) with and (b) without shot transition

2.2 (Audio)Visual Signal Segmentation

Visual segmentation of the visual part of the multimedia data is currently mainly used for indexing audio-visual data (video recordings). During visual segmentation, the detection of changes in the more or less homogeneous parts of the visual signal is solved, where the homogeneous part is understood as a sequence of video frames in which the information does not considerably change. The location of the detected change is the segment boundary (shot transition), which is normally marked during video editing [1,7]. Various video effects are also occasionally used when creating a video recording – e.g. merging single images or creating gradual transitions between images [5,6]. Research in this area in recent years has been primarily focused on finding the boundaries of visual segments in areas where complex computer-generated transition effects have been used. Nowadays, the state-of-the-art in the area of visual signal segmentation (Shot Transition Detection - STD) are the techniques based on DNNs or CNNs [18].

Our STD system is based on CNN. A special data set from our TV news video archive was prepared for training and evaluation of CNN. Shot transitions from 3 h of TV news were manually selected. It was 2001 shot transitions in total. Video recordings from dataset were segmented into segments where six consecutive video frames were in one new image. The shot transitions was in the middle, see Fig. 3. So, 4002 images were prepared (2001 with shot transitions and 2001 without). There were used 80% of images for training of CNN and 20% of images for evaluation. The images were resized (192 × 1536 pixels) and parameterized by Histogram of Oriented Gradients (HOG) with cell (64 × 64) in the first step. The CNN was composed of three convolutional and one fully connected layers. Kernels with size 8 × 8 was utilized in convolutional layers and fully connected layer had 4096 neurons. Softmax was used as activation function in output layer and ReLU in other layers. CNN was trained in 5 epochs, see Table 2.

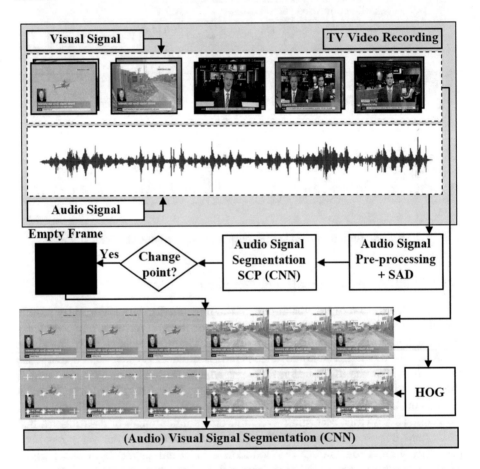

Fig. 4. A principle of audio-visual TV broadcast signal segmentation

The main idea of this work was to use information from audio SCP for improving of visual segmentation. Different ways of integration have been tested, especially at the architecture level of CNN. Finally, the easiest and the most reliable way was to add empty (black) frame on position in video signal where change point from audio SCP was computed, see Fig. 4.

Visual Segmentation Rate (VSR) was used as a metric of the performance of (audio)-visual segmentation:

$$VSR = \frac{H - I}{N} * 100[\%] \tag{1}$$

Table 2. CNN architecture for (audio)visual signal segmentation

Layer	Architecture	Transfer function
Convolution	16 kernels 8 × 8	ReLU
Convolution	32 kernels 8 × 8	ReLU
Convolution	64 kernels 8 × 8	ReLU
Fully connected	4096 neurons	ReLU
Output	2 neurons	SoftMax

where N is the number of all shot transitions, H is the number of correctly identified shot transitions and I is the number of insertions.

VSR for visual segmentation only was 88,6% and it was 90,5% if information from audio SCP was integrated.

3 Conclusion

A novel method for audio-visual TV broadcast signal segmentation is presented and described in this paper. It is shown that audio part of signal segmentation can improve results from visual signal segmentation mainly in TV news video recordings.

We would like to increase the dataset for training of CNN for audio-visual signal segmentation and re-train CNN in the near future. After this step, we plan to integrate resulting audio-visual TV broadcast signal segmentation into our prototype of TV broadcast transcription.

Acknowledgements. The research was supported by the Technology Agency of the Czech Republic in project no. TH03010018.

References

1. Clua, E., Fonseca, M., Conci, A., Montenegro, A.: Detecting shot transitions based on video content. In: 2008 15th International Conference on Systems, Signals and Image Processing, pp. 339–342. IEEE (2008)
2. Dahl, G.E., Sainath, T.N., Hinton, G.E.: Improving deep neural networks for LVCSR using rectified linear units and dropout. In: 2013 IEEE International Conference on Acoustics, Speech and Signal Processing, pp. 8609–8613. IEEE (2013)
3. Deng, L., Li, J., Huang, J.T., Yao, K., Yu, D., Seide, F., Seltzer, M., Zweig, G., He, X., Williams, J., et al.: Recent advances in deep learning for speech research at microsoft. In: 2013 IEEE International Conference on Acoustics, Speech and Signal Processing, pp. 8604–8608. IEEE (2013)
4. Hrúz, M., Zajíc, Z.: Convolutional neural network for speaker change detection in telephone speaker diarization system. In: 2017 IEEE International Conference on Acoustics, Speech and Signal Processing (ICASSP), pp. 4945–4949. IEEE (2017)

5. Huang, M., Jiang, J., Liang, S., Zhang, Y.: A shot gradual changes detection algorithm based on adjacent scale wavelet transform. In: PIAGENG 2009: Intelligent Information, Control, and Communication Technology for Agricultural Engineering, vol. 7490, p. 74901F. International Society for Optics and Photonics (2009)
6. Kim, W.H., Jeong, T.I., Kim, J.N.: Video segmentation algorithm using threshold and weighting based on moving sliding window. In: 2009 11th International Conference on Advanced Communication Technology, vol. 3, pp. 1781–1784. IEEE (2009)
7. Lefèvre, S., Holler, J., Vincent, N.: A review of real-time segmentation of uncompressed video sequences for content-based search and retrieval. Real-Time Imaging 9(1), 73–98 (2003)
8. Mateju, L., Cerva, P., Zdansky, J., Malek, J.: Speech activity detection in online broadcast transcription using deep neural networks and weighted finite state transducers. In: 2017 IEEE International Conference on Acoustics, Speech and Signal Processing (ICASSP), pp. 5460–5464. IEEE (2017)
9. Neri, L., Pinheiro, H., Ren, T., Cavalcanti, G., Adami, A.: Speaker segmentation using i-vectors in meeting domain. In: ICASSP 2017, pp. 5455–5459. IEEE (2017)
10. Nouza, J., Blavka, K., Zdansky, J., Cerva, P., Silovsky, J., Bohac, M., Chaloupka, J., Kucharova, M., Seps, L.: Large-scale processing, indexing and search system for czech audio-visual cultural heritage archives. In: 2012 IEEE 14th International Workshop on Multimedia Signal Processing (MMSP), pp. 337–342. IEEE (2012)
11. Nouza, J., Cerva, P., Safarik, R.: Cross-lingual adaptation of broadcast transcription system to polish language using public data sources. In: Language and Technology Conference, pp. 31–41. Springer (2015)
12. Nouza, J., Safarik, R., Cerva, P.: ASR for South Slavic languages developed in almost automated way. In: INTERSPEECH, pp. 3868–3872 (2016)
13. Nouza, J., Silovsky, J., Zdansky, J., Cerva, P., Kroul, M., Chaloupka, J.: Czech-to-Slovak adapted broadcast news transcription system. In: Ninth Annual Conference of the International Speech Communication Association (2008)
14. Nouza, J., Zdánský, J., David, P., Cerva, P., Kolorenc, J., Nejedlová, D.: Fully automated system for czech spoken broadcast transcription with very large (300k+) lexicon. In: Ninth European Conference on Speech Communication and Technology (2005)
15. Rouvier, M., Dupuy, G., Gay, P., Khoury, E., Merlin, T., Meignier, S.: An open-source state-of-the-art toolbox for broadcast news diarization. In: Interspeech (2013)
16. Sarkar, A., Dasgupta, S., Naskar, S., Bandyopadhyay, S.: Says who? deep learning models for joint speech recognition, segmentation and diarization. In: ICASSP 2018, pp. 5229–5233. IEEE (2018)
17. Sinha, R., Tranter, S.E., Gales, M.J., Woodland, P.C.: The Cambridge university march 2005 speaker diarisation system. In: Ninth European Conference on Speech Communication and Technology (2005)
18. Tong, W., Song, L., Yang, X., Qu, H., Xie, R.: CNN-based shot boundary detection and video annotation. In: 2015 IEEE International Symposium on Broadband Multimedia Systems and Broadcasting, pp. 1–5. IEEE (2015)
19. Vandecatseye, A., Martens, J.P., Neto, J.P., Meinedo, H., Garcia-Mateo, C., Dieguez-Tirado, J., Mihelic, F., Zibert, J., Nouza, J., David, P., et al.: The cost 278 pan-European broadcast news database. In: LREC (2004)

Optimizing Training Data and Hyperparameters of Support Vector Machines Using a Memetic Algorithm

Wojciech Dudzik$^{(\boxtimes)}$, Michal Kawulok, and Jakub Nalepa

Silesian University of Technology, Gliwice, Poland
{wojciech.dudzik,michal.kawulok,jakub.nalepa}@polsl.pl

Abstract. Support vector machine (SVM) is a well-known machine learning algorithm widely used for classification and regression problems. Despite the high prediction rate of this technique in a wide range of real applications, the efficiency of SVM and its classification performance highly depends on the hyperparameters setting as well as the selection of feature subset. Moreover, high memory and computational complexity of SVM training can be a limiting factor for its application on huge dataset. In this work we propose a novel memetic algorithm for support vector machine called SE-SVM to address mentioned problems. We use evolutionary techniques that optimize hyperparameters and select features and training set simultaneously. The algorithm is applied to seven datasets. All of that datasets are binary classification problem. We compare the SE-SVM to different evolutionary algorithms, random search techniques and other well-established classifiers. The experimental results show that the end result obtained by SE-SVM achieves high classification performance with a shorter training and classification time.

Keywords: Evolutionary algorithms · Support vector machine

1 Introduction

SVMs showed state-of-the-art performance in real-world applications such as text categorization, hand-written character recognition, image classification and others. Nonetheless, using SVM is hard due to its sensitivity to hyperparameters ($\mathcal{M}$). The need for optimizing these hyperparameters is undeniable, as classification efficiency can change vastly depending on their values. A high computational cost of a grid search (GS) algorithm is a trigger for development of new optimization algorithms. In this work we focus on non-linear binary classification task. To obtain a non-linearity with SVM we have to use a kernel function. These functions have their own hyperparameters that need to be tuned. Here we are using radial basis function as kernel. Considering this there are only two hyperparameters C—slack variable of SVM and γ—width of RBF kernel.

The next problem of expensive training can be tackled with proper selection of training set (T). Using a subset of T (T') can be sufficient, as only a small part

© Springer Nature Switzerland AG 2020
A. Gruca et al. (Eds.): ICMMI 2019, AISC 1061, pp. 229–238, 2020.
https://doi.org/10.1007/978-3-030-31964-9_22

of T is chosen as support vectors (S). As computational complexity depends on the size of training set, with $O(\|T^2\|)$ memory complexity and time complexity of $O(\|T^3\|)$, there is an urgent need to reduce this size as much as possible without losing classification performance.

The last problem is selection of a feature set (F). We cannot *a-priori* know which features are useful for a classifier. Redundant features can adversely affect classification accuracy and time performance. At the same time, too few features can lower accuracy, as there may not be enough information provided.

Furthermore, there exists mutual dependence of selected $\mathcal{M}$ on choice of T' and F' and vice versa. This dependence makes optimization of $\mathcal{M}$, T' and F' not trivial. All that problems may tend to drop the use of SVM classifier as considerably hard to tune. The key aspects of our contribution are as follows: (i) novel memetic algorithm called Simultaneously tuned SVM (SE-SVM) that is the first to optimize $\mathcal{M}$, T' and F' at once, (ii) design of new dynamically-sized chromosome, (iii) possibility to use expert knowledge for initial feature selection.

In next section we will outline current methods for optimizing SVM present in literature. In Sect. 3 we describe our approach. Section 4 presents results of experiments and provides discussion. In Sect. 5 the paper is concluded.

2   Literature Review

Optimization of different aspects of SVMs is an active research topic. It can be divided into several categories: (i) optimizing hyperparameters, (ii) optimizing T, (iii) optimizing F, (iv) combining any of these aspects. For all of these problems evolutionary and bio-inspired methods were used numerous times but it seems that combining all of these techniques was not thoroughly explored yet.

In the field of hyperparameter optimization, a lot of work is put in evolutionary algorithms (EA) such as: particle swarm optimization [16] and genetic algorithms [2]. In another recent work, the SVM hyperparameters are optimized by fast messy GA. One more evolutionary strategy proposes a bat algorithm that outperforms previously mentioned methods [14].

The methods for optimizing T are summed up in a review [11] which summarize current work in the field. Methods used for T' selection can be divided into two groups: (i) the ones whose complexity is dependent on the cardinality of T, and (ii) those that are independent of T size. The algorithms from the first group exploit the information about the layout of T, including clustering method [13], exploiting statistical properties of T [5], and neighbourhood analysis [6]. The main drawback of those methods is trading SVM time and memory complexity to those of other algorithms. The second group includes random sampling and EA, where the second group gets more attention. Random sampling is exploited in [1], where a random T' is drawn and refined iteratively. In EA there is a highly effective memetic algorithm to select training data for SVM (MASVM) [10]. Other work includes evolutionary wrapper technique for T' selection [15] and our previous work called alternating genetic algorithm (abbreviated as ALGA) for optimizing the SVM model alongside SVM training sets [8]. We observed that different SVM models ($\mathcal{M}$) may be optimal for different T.

In most cases, selecting feature subset $(\boldsymbol{F'})$ for SVM is coupled with hyper-parameters optimization [7,9]. One of the latest works includes feature weighting [12] which scales features according to their relevance. In our previous work we proposed feature selection followed by alternating memetic algorithm (FSALMA) presented in [3], where we optimize $\mathcal{M}$ and $\boldsymbol{T'}$ on preselected $\boldsymbol{F}$. Although much work has been done in the field, combined techniques for all aspects have not yet been thoroughly tested.

3   Proposed Method

In the Fig. 1 we present the general schema for our evolution process. The operations marked with asterisk are performed only for $\boldsymbol{T}$ and $\boldsymbol{F}$ part of chromosome which is described in Sect. 3.1. First, our workflow starts with ranking $\boldsymbol{F}$ averaging results of mutual information, variance thresholding, recursive feature elimination and stability selection algorithms. These scores are used as probabilities in roulette wheel initialization of feature sets, where $K_F = 4$ features are selected. On initialization only top 20% of features are considered for getting into chromosome. Initialization of $\boldsymbol{T'}$ starts with choosing a small random subset of $\boldsymbol{T}$ in size of $K_T = 16$ for each of classes. This means that our $\boldsymbol{T'}$ is always perfectly balanced. The $\mathcal{M}$ is initialized deterministically with a logarithmic step in range $[10^{-5}, 10^3]$. If population cannot be divided into a grid (square root is not an integer), the rest of population is drawn randomly from this interval. At the end we evaluate created population with the scheme presented in Sect. 3.2.

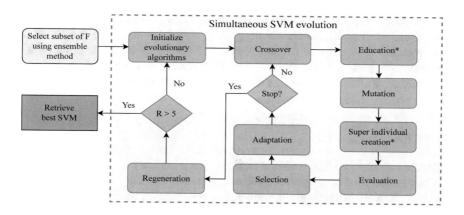

Fig. 1. Flowchart of SE-SVM.

Starting optimization in each iteration, $Q = 20$ individuals (I_i) are created by the crossover process. This consists of three different operators for each part of chromosome. The hyperparameter values of a new child become $x_{a+b} = x_a + \alpha \cdot (x_a - x_b)$, where α is the weight randomly drawn from the interval $[-0.5, 1.5]$. When it comes to $\boldsymbol{T'}$ and $\boldsymbol{F'}$ part, two sets (of features or training examples) are summed. As summation of sets may lead to uncontrolled growth in size of these sets, we randomly select no more then K_T (or K_F) elements.

After crossover for both parts of T' and F', we utilize memetic operation of local education where we exploit information about previous individuals to enhance current solutions. For training set selection we create a pool of vectors that were chosen as support vectors (S) in previous generation. The education operation is performed with $P_{Te} = 30\%$ probability and exchange randomly selected $S_e = 20\%$ of vectors to the ones that are in pool. If pool is too small or there is lack of vectors to maintain class balance the operation is ended with maximum possible number of exchanges. For F' we create a pool of features that are associated with high ranked individuals. This process takes the whole population and create the histogram of features from 50% best individuals. Next, only features that occur more than once are left there. The education is performed in the same way as for T' with same parameter settings.

Next, each individual is mutated with probability of P_m for $\mathcal{M}$, $P_t = 30\%$ for T' and $P_f = 50\%$ for F'. Mutation of $\mathcal{M}$ consists in modifying a value x (C or γ) within a range $x \in [x - u \cdot x, x + u \cdot x]$, where u is randomly drawn from $[0.0, 0.1]$. Mutation for T' consists in replacing $f_t = 20\%$ of examples to random ones taken from T. For F', $f_f = 10\%$ features are replaced to ones that are above mean in ranking used for initialization.

Another memetic operation is super individual creation. These individuals are created from pools of features or S in random manner. We select up to $K_T \cdot N_c$ (N_c—number of classes) or K_F elements for those individuals. For $\mathcal{M}$ we take the best individual and match it with those super individuals of T' and F'. Next step is evaluating newly created individuals and calculating their fitness.

Eventually, we select the Q fittest individuals to maintain steady size of population. At the end, we adaptively grow K_T and K_F based on current fitness improvement. The process of evolution lasts till there is no improvement in the average fitness of population. If no improvement can be made, we perform regeneration of all individuals. After $R = 5$ regenerations, we take the best model from the entire evolutionary process.

3.1   Chromosome Design

Chromosome of SE-SVM consists of three parts as shown on Fig. 2. The first part is composed of hyperparameters and kernel type. The design can hold any arbitrary number of them giving selected kernel type. This type is not changed during evolution so this part has constant size. Next parts are two sets, one for training examples and one for features. We did not use classical binary encoding for those sets as this is impractical for big datasets. Instead, we are holding only the IDs of currently selected elements. These IDs are numbers of rows and columns to be selected. We maintain a uniqueness of those IDs through whole evolution process. Moreover, these sets grow during the evolution process starting from some arbitrary small value of K_F for F and K_T for T. This makes the size of the chromosome dynamic and variable during evolution.

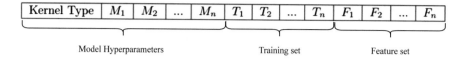

Fig. 2. Chromosome design of SE-SVM.

3.2 Fitness Calculation

For measuring fitness, we applied area under the receiver operating characteristic curve (AUC) as metric. The process of fitness calculation is presented in the Fig. 3. As we optimize both the training and the feature set we need to filter out original training set selecting only those elements that are in a chromosome. This is presented as choosing only intersected elements of T' and F'. Next, we apply it together with current hyperparameters and train SVM model. Later, this model classifies validation set which is filtered only by features. As the last step, we calculate AUC metric on scores obtained for V.

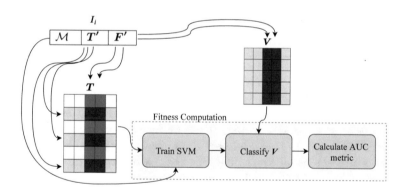

Fig. 3. Process of computing fintess for single individual of SE-SVM.

4   Experiments and Discussion

All of the experiments were performed on Intel i5-6500 CPU with 16 GB RAM. We are using OpenCV 3.3.1 implementation of SVM algorithm. Datasets were divided into five folds containing training, validation and test set in 3:1:1 proportion, respectively. There was no overlap between subsets, this means that $V \cap \Psi \cap T = \varnothing$. We picked seven binary datasets that are presented in Table 1.

The presented results are average scores over all five folds. Each algorithm (except grid search) was run 10 times per fold. The best AUC scores and times are boldfaced in tables (excluding random search techniques). Feature ranking

Table 1. Benchmark datasets from UCI and Kaggle repositories.

Dataset	Size	Features	Class balance (T+/T-)
Ionosphere	351	34	0.64
Madelon	2600	500	1.00
Gisette	7000	5000	1.00
German	1000	24	0.30
Wisconsin	699	9	0.36
Default or Credit Card Clients (DCCC)	30000	23	0.22
Credit Card Fraud Detection (CCFD)	284807	30	0.0017

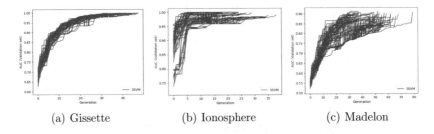

(a) Gissette   (b) Ionosphere   (c) Madelon

Fig. 4. Overview of fitness growth during SE-SVM evolution for selected datasets.

performed before starting EA is not included into time measurements. Before running GS we applied recursive feature elimination described in [3].

In Fig. 4 we can see the growth of fitness in selected datasets for all runs across different folds. As it can be seen the repeated high and constant growth of fitness is present in all cases. The flat regions of fitness for the last 10–15 generations seen in Fig. 4(a) and (b) are due to regeneration technique.

In Table 2 we can see all of the results gathered together for all datasets. Looking at the training times we achieved very consistent results (smaller standard deviation in comparison to others) although not always getting the best time. For the bigger datasets we clearly outperformed GS and ESVM algorithms. In DCCC and CCFD the FSALMA was faster but it does not perform feature selection along the way. Also on DCCC it seems to stop too early. When it comes to fitness values we are not worse than 0.010 AUC in comparison to the best algorithm for most datasets, regarding both V and Ψ. Moreover, we outperformed other EA in all cases. The only exception to that is German dataset where too aggressive feature selection and early stopping causes poor results. However, the fitness on Ψ was still second best. It is worth noting that on 3 out of 7 times we get the smallest number of S. Considering this and ability to reduce the feature count, listed in last column, we are able to achieve fast training and classification times performance with getting good generalization. In the cases where we did not get the smallest number of S, we outperformed these classifiers looking at the fitness values for Ψ. The results on CCFD dataset show that high imbalance

Table 2. Results of experiments on selected datasets. Values are averaged scores from all runs followed by standard deviation

Set	Algorithm	Training time [s]	Fit. V	S	Fit. Ψ	Classification time V [ms]	Features
Ionosphere	**SE-SVM**	0.34 ± 0.14	.991 ± .01	61 ± 20	.961 ± .03	0.11 ± 0.04	9.4 ± 2
	ESVM	0.82 ± 0.45	.995 ± .01	51 ± 22	.950 ± .04	**0.08** ± 0.03	8.6 ± 3
	FSALMA	0.34 ± 0.15	**.996** ± .01	**41** ± 21	.953 ± .04	0.12 ± 0.06	24.4 ± 9
	GS	**0.25** ± 0.06	.983 ± .02	125 ± 40	**.968** ± .04	0.21 ± 0.06	24.4 ± 10
	RS	0.46 ± 0.31	.987 ± .01	135 ± 39	.963 ± .03	0.28 ± 0.12	19 ± 8
	RSEH	0.07 ± 0.03	.979 ± .02	55 ± 12	.947 ± .03	0.11 ± 0.03	8.8 ± 1
	RSIP	0.04 ± 0.01	.949 ± .02	23 ± 8	.886 ± .06	0.05 ± 0.01	4 ± 0
Madelon	**SE-SVM**	118.1 ± 30.2	**.893** ± .03	**724** ± 227	**.868** ± .03	**6.83** ± 2.33	18.5 ± 3
	ESVM	215.1 ± 154.3	.890 ± .05	740 ± 240	.861 ± .05	7 ± 3.44	17.3 ± 5
	FSALMA	87.7 ± 15.8	.855 ± .04	982 ± 78	.814 ± .04	21.23 ± 4.08	50 ± 0
	GS	**23.3** ± 1	.832 ± .04	1514 ± 27	.828 ± .03	29.92 ± 0.42	50 ± 0
	RS	100.9 ± 14.5	.597 ± .04	909 ± 387	.569 ± .06	61.4 ± 56.79	166.2 ± 150
	RSEH	11.5 ± 3	.595 ± .02	660 ± 113	.564 ± .03	6.51 ± 1.33	19.8 ± 2
	RSIP	1.4 ± 0.2	.573 ± .01	31 ± 3	.501 ± .02	0.5 ± 0.1	4 ± 0
Gissette	**SE-SVM**	**222** ± 126	.995 ± .00	**440** ± 195	.994 ± .00	**108** ± 74	225 ± 117
	ESVM	389 ± 499	.996 ± .00	463 ± 273	.995 ± .00	175 ± 144	363 ± 169
	FSALMA	358 ± 446	.995 ± .00	534 ± 222	.993 ± .00	817 ± 779	1300 ± 990
	GS	2938 ± 2842	**.997** ± .00	1331 ± 237	**.997** ± .00	1888 ± 1826	1300 ± 1095
	RS	8061 ± 331	.713 ± .10	1890 ± 1028	.717 ± .10	273 ± 810	116 ± 294
	RSEH	55 ± 9	.771 ± .16	410 ± 68	.771 ± .16	83 ± 13	138 ± 16
	RSIP	30 ± 1	.780 ± .02	30 ± 4	.772 ± .02	1.24 ± 0.24	4 ± 0
Wisconsin	**SE-SVM**	**0.07** ± 0.03	.997 ± .00	23 ± 16	.987 ± .02	**0.11** ± 0.03	6.1 ± 2
	ESVM	0.21 ± 0.1	.998 ± .00	22 ± 14	.988 ± .02	0.11 ± 0.04	5.4 ± 2
	FSALMA	0.11 ± 0.04	**.998** ± .00	**18** ± 15	.986 ± .02	0.11 ± 0.04	6.8 ± 2
	GS	0.32 ± 0.1	.995 ± .00	163 ± 122	**.995** ± .00	0.28 ± 0.15	6.8 ± 2
	RS	0.09 ± 0.04	.994 ± .01	116 ± 66	.971 ± .03	0.32 ± 0.15	5.4 ± 2
	RSEH	0.02 ± 0.01	.992 ± .01	17 ± 5	.976 ± .03	0.12 ± 0.02	5.8 ± 1
	RSIP	0.02 ± 0.02	.995 ± .00	25 ± 10	.989 ± .01	0.11 ± 0.05	4 ± 0
German	**SE-SVM**	**0.6** ± 0.4	.799 ± .03	87 ± 51	.747 ± .05	**0.32** ± 0.17	9.1 ± 2
	ESVM	14.5 ± 16.7	**.818** ± .03	136 ± 45	.738 ± .04	0.44 ± 0.24	7.5 ± 2
	FSALMA	1.5 ± 1	.814 ± .03	115 ± 46	.742 ± .06	0.42 ± 0.16	12.4 ± 5
	GS	1.4 ± 0.4	.774 ± .03	355 ± 10	**.775** ± .04	1.16 ± 0.47	12.4 ± 6
	RS	7.4 ± 6.4	.727 ± .03	293 ± 147	.642 ± .06	1.34 ± 0.95	10.4 ± 6
	RSEH	0.1 ± 0	.721 ± .03	80 ± 12	.631 ± .04	0.38 ± 0.06	8.6 ± 1
	RSIP	0.1 ± 0	.720 ± .03	29 ± 4	.679 ± .02	0.15 ± 0.01	4 ± 0
DCCC	**SE-SVM**	36 ± 21	.759 ± .02	693 ± 547	**.752** ± .02	69.5 ± 50.5	15.5 ± 1.7
	ESVM	105 ± 110	**.760** ± .02	776 ± 831	.748 ± .02	83.2 ± 91.6	13.8 ± 3.9
	FSALMA	9 ± 12	.750 ± .02	**261** ± 323	.737 ± .02	**32.1** ± 35.0	18.6 ± 3.0
	GS	7567 ± 3435	.723 ± .03	7791 ± 510	.727 ± .03	770.6 ± 118.9	18.6 ± 3.3
	RSEH	44 ± 16	.684 ± .01	373 ± 131	.683 ± .02	47.6 ± 26.1	15.0 ± 0.7
	RSIP	3 ± 1	.715 ± .02	29 ± 6	.712 ± .02	12.7 ± 7.5	4.0 ± 0.0
CCFD	**SE-SVM**	81 ± 15	.989 ± .01	30 ± 22	**.975** ± .04	71.7 ± 32.0	16.9 ± 5.4
	ESVM	110 ± 32	**.991** ± .01	29 ± 18	.966 ± .03	68.6 ± 24.6	16.4 ± 5.1
	FSALMA	19 ± 9	.983 ± .01	29 ± 19	.964 ± .03	**66.5** ± 26.6	16.4 ± 10.8
	GS	1578 ± 435	.940 ± .04	446 ± 136	.928 ± .04	329.2 ± 53.1	9.6 ± 7.6
	RSEH	30 ± 13	.974 ± .02	18 ± 11	.918 ± .08	75.6 ± 21.7	15.2 ± 3
	RSIP	21 ± 7	.976 ± .01	18 ± 10	.963 ± .02	56.3 ± 14.7	4.0 ± 0

is not a problem for SE-SVM. More detailed comparison with random search techniques is described in Sect. 4.1. In Sect. 4.2 we analyze statistical difference of SE-SVM and in Sect. 4.3 we compare ourselves to other classifiers.

4.1 Comparison to Random Search

We incorporated three different techniques for random search: (i) random search where feature and training set size were chosen randomly (RS) (ii) random search with feature and training set size taken as per fold averages from end results of SE-SVM (RSEH) (iii) random search composed of initial populations from SE-SVM (RSIP). These techniques were used to check whether our approach is better than random search with respect of using the same number of evaluations performed for both techniques. The results presented in Table 2 show that we always achieve higher fitness scores although random search techniques performed surprisingly well on small datasets, especially Ionosphere. For DCCC and CCFD datasets RS took too long to complete so there are no results presented.

4.2 Statistical Difference Test

We performed Wilcoxon test to verify whether there are statistically significant differences between the results obtained with SE-SVM and other tested algorithms. As many ideas exploited in our previous evolutionarily tuned SVM (ESVM) [4] algorithm are used for SE-SVM we were expecting that there might be no statistical difference between those two algorithms. Our expectations have not been confirmed. Not including several exceptions ($p > 0.05$ which are bold-faced), the presented method is statistically different from all compared algorithms as shown in Table 3, where most of results show $p < 0.001$.

Table 3. Wilcoxon signed-rank test comparing results of SE-SVM to that of other techniques for given datasets. The values represent $p - value$.

Dataset	ESVM	FSALMA	GS	RS	RSEH	RSIP
Ionosphere	<0.001	<0.001	<0.001	<0.001	<0.001	<0.001
Madelon	**>0.2**	<0.001	<0.001	<0.001	<0.001	<0.001
Gissette	<0.001	**>0.05**	<0.001	<0.001	<0.001	<0.001
Wisconsin	<0.001	**>0.2**	<0.001	<0.001	<0.001	<0.001
German	<0.001	<0.001	<0.001	<0.001	<0.001	<0.001
DCCC	<0.001	<0.001	<0.001	–	<0.001	<0.001
CCFD	<0.001	<0.001	<0.001	–	<0.001	<0.001

4.3   Comparison to Other Classifiers

In Table 4, we compared our SE-SVM with other popular classifiers. For KNN we used $K = 5$ and for ExtraTree the number of trees was set to $N = 100$. In case of Lasso and Logistic Regression, we used cross-validation for hyperparameter tuning. As it can be seen we are not always the best among other techniques but overall the scores are close to the best results. On the Ionosphere we found that the small size of V might be problematic as in some cases we overfitted to it. Also the division into V and T might not be representative for Ψ in all of the folds. That is why ensemble method of ExtraTree could achieve better average score. The greatest difference between SE-SVM and other method can be seen on Madelon dataset where we outperformed them as proper feature selection is crucial here. Only 20 from 500 features are informative. On the Gisette and Wisconsin datasets all of presented methods achieved very high (almost perfect) AUC scores. As Wisconsin is relatively small it is hard to make better generalization without overfitting. The biggest gap between SE-SVM and other methods can be seen on German dataset. Its poor performance in this case can be explained by too aggressive feature selection. As we use forward feature selection we rarely select all of the features which in case of small number of features can limit the final performance. On the other hand, this problem is not seen on DCCC and CCFD datasets. One more problem is poor fold division as performance on V does not represent the results obtained on Ψ. Although we got the best score on DCCC we won only by small margin with ExtraTree.

Table 4. Comparison of SE-SVM with other classifiers. The results are AUC on Test set averaged over 5-folds.

Algorithm↓ Set →	Ionosphere	Madelon	Gisette	Wisconsin	German	DCCC	CCFD
ExtraTree	**0.987**	0.698	**0.996**	0.994	0.782	0.747	0.928
Lasso	0.907	0.640	**0.996**	0.994	0.791	0.715	**0.975**
KNN	0.920	0.581	0.988	0.990	0.699	0.699	0.854
Logistic Regression	0.918	0.630	**0.996**	**0.995**	**0.797**	0.721	**0.975**
SE-SVM	0.961	**0.868**	0.994	0.988	0.747	**0.752**	**0.975**

5   Conclusions

The main novelty of presented memetic algorithm is simultaneously optimization of all three aspects $\mathcal{M}$, T' and F'. We performed analysis with comparison to other EA, random search techniques and GS as a baseline. As showed SE-SVM achieves good results while being able to reduce classification time greatly. Furthermore we performed statistical analysis of presented result showing that our approach is statistically different in most cases. Although not giving the best result in comparison with other classifiers on all presented datasets its performance was competitive to these methods as AUC scores did not differ greatly.

It is worth noting that we achieved much faster training time in comparison to ESVM which is the only technique that optimizes $\mathcal{M}$, $\boldsymbol{T'}$ and $\boldsymbol{F'}$. In future we plan to enhance our method with better evolutionary operators and remove the regeneration operation.

Acknowledgements. This work was supported by the National Science Centre under Grant DEC-2017/25/B/ST6/00474. W. Dudzik was co-financed by the European Union through the European Social Fund (grant POWR.03.02.00-00-I029) and by the Silesian University of Technology (02/020/BKM18/0155).

References

1. Balcázar, J., Dai, Y., Watanabe, O.: A random sampling technique for training support vector machines. In: ALT, pp. 119–134. Springer, Heidelberg (2001)
2. Chou, J.S., Cheng, M.Y., Wu, Y.W., Pham, A.D.: Optimizing parameters of support vector machine using fast messy genetic algorithm for dispute classification. Expert Syst. Appl. **41**(8), 3955–3964 (2014)
3. Dudzik, W., Nalepa, J., Kawulok, M.: Automated optimization of non-linear SVMs for binary classification. In: InCoS, pp. 504–513. Springer, Cham (2018)
4. Dudzik, W., Nalepa, J., Kawulok, M.: Evolutionarily tuned support vector machines. In: GECCO 2019 Companion. ACM, Prague (2019)
5. Guo, L., Boukir, S.: Fast data selection for SVM training using ensemble margin. Pattern Recogn. Lett. **51**, 112–119 (2015)
6. He, Q., Xie, Z., Hu, Q., Wu, C.: Neighborhood based sample and feature selection for svm classification learning. Neurocomput. **74**(10), 1585–1594 (2011)
7. Huang, C.L., Wang, C.J.: A GA-based feature selection and parameters optimizationfor SVMs. Expert Syst. Appl. **31**(2), 231–240 (2006)
8. Kawulok, M., Nalepa, J., Dudzik, W.: An alternating genetic algorithm for selecting SVM model and training set. In: MCPR, pp. 94–104. Springer, Cham (2017)
9. Lin, S.W., Ying, K.C., Chen, S.C., Lee, Z.J.: Particle swarm optimization for parameter determination and feature selection of support vector machines. Expert Syst. Appl. **35**(4), 1817–1824 (2008)
10. Nalepa, J., Kawulok, M.: A memetic algorithm to select training data for support vector machines. In: GECCO 2014, pp. 573–580. ACM, New York (2014)
11. Nalepa, J., Kawulok, M.: Selecting training sets for support vector machines: a review. Artif. Intell. Rev. **52**, 1–44 (2018)
12. Phan, A.V., Nguyen, M.L., Bui, L.T.: Feature weighting and SVM parameters optimization based on genetic algorithms for classification problems. Appl. Intell. **46**(2), 455–469 (2017)
13. Shen, X.J., Mu, L., Li, Z., Wu, H.X., Gou, J.P., Chen, X.: Large-scale SVM classification with redundant data reduction. Neurocomputing **172**, 189–197 (2016)
14. Tharwat, A., Hassanien, A.E., Elnaghi, B.E.: A BA-based algorithm for parameter optimization of support vector machine. Pattern Recogn. Lett. **93**, 13–22 (2017)
15. Verbiest, N., Derrac, J., Cornelis, C., García, S., Herrera, F.: Evolutionary wrapper approaches for training set selection as preprocessing mechanism for SVMs. Appl. Soft Comput. **38**(C), 10–22 (2016)
16. Zhang, X., Qiu, D., Chen, F.: Support vector machine with parameter optimization by a novel hybrid method and its application to fault diagnosis. Neurocomputing **149**, 641–651 (2015)

Induction of Centre-Based Biclusters
in Terms of Boolean Reasoning

Marcin Michalak[(⊠)]

Institute of Informatics, Silesian University of Technology,
ul. Akademicka 16, 44-100 Gliwice, Poland
Marcin.Michalak@polsl.pl

Abstract. Biclustering is considered as the method of finding two-dimensional subgroups in a matrix of scalars. The paper introduces a new approach to biclustering continuous matrices on the basis of boolean function analysis, where it is assumed that found bicluster must fulfill two criteria: it should cover the specified value called a centre and the maximal difference between the mentioned centre and cells of bicluster should not exceed an assumed margin. The paper also presents mathematical proofs of the correctness of such approach. The application of new approach for biclustering was applied on artificial data.

Keywords: Biclustering · Boolean reasoning ·
Continuous biclustering · Biclusters of similarity

1 Introduction

Biclustering (called also co-clustering, two-dimensional clustering, two-mode clustering) is the method of twodimensional data analysis started in 1970's by Hartigan [4]. Since then it has been successfully used in gene expression analysis [12], text mining [3,11], data exploration [6] and many other applications [2].

The notion of biclustering is very often mistaken with clustering. Multidimensional clustering (such k-means or DBSCAN) introduces a partitions of set of objects from the mathematical point of view (in case of DBSCAN or any other clustering techniques that allows to get some nonclustered objects considered as the "noise" in the data, such "noise" is just another element of the clustering results). In case of biclustering we are interested in finding some submatrices of the given one, whose elements fulfill the defined condition of similarity. On the one hand we may be interested in exact biclustering of the data, where all elements in biclusters have the same value (it is common for binary or discrete value matrices). On the other hand we may be interested in finding biclusters, whose elements are not equal but for example the maximal absolute difference between bicluster elements does not exceed an assumed level. The difference between clustering and biclustering is shown in Figs. 1 and 2.

In Fig. 1 the possible results of two-dimensional data is presented, where clustering of objects (left) and clustering of features (right) was applied. In Fig. 2

© Springer Nature Switzerland AG 2020
A. Gruca et al. (Eds.): ICMMI 2019, AISC 1061, pp. 239–248, 2020.
https://doi.org/10.1007/978-3-030-31964-9_23

	f1	f2	f3	f4
o1	1	2	30	40
o2	2	3	31	41
o3	10	11	-5	-8
o4	11	12	-4	-7
o5	101	102	103	104

	f1	f2	f3	f4
o1	31	32	88	91
o2	1	4	87	95
o3	2	4	90	93
o4	50	52	91	94
o5	50	53	90	94

Fig. 1. Possible results of object (left) and feature (right) clustering.

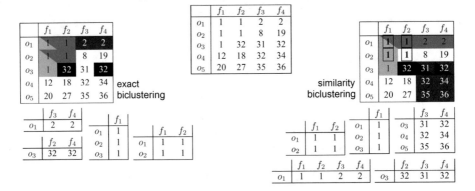

Fig. 2. Possible results of discrete data biclustering: the original data are presented in the upper centre, results of exact biclustering are presented on the left side while results of similarity based biclustering (here: the maximal difference between bicluster elements is not higher than 5) are presented on the right side; in both cases only non-trivial (not single cell) biclusters are mentioned.

we may see the possible results of exact (left) and similarity (right) biclustering. As we can see in the second case (similarity biclustering) the obtained results don't have to cover all the matrix at all (trivial biclusters that contain only one cell are not taken into consideration).

In the paper the new approach for bicluster induction is presented. It is developed from the boolean reasoning paradigm [1,10]. Foundations of boolean reasoning in biclustering for discrete and boolean matrices were successfully presented in [8], afterwards it was extended for matrices of continuous values [9]. In the second paper two kinds of continuous biclusters were defined: bicluster of similarity and bicluster of dissimilarity. Bicluster of similarity denotes the submatrix of the given matrix, whose elements do not differ from themselves by more than assumed level. On the other hand the bicluster of dissimilarity (or in other words—the bicluster of chaos) contains only such elements that differ from each other at least by the specified value. The mentioned two papers provide not only the definition of biclusters and the algorithm of their induction in terms of boolean reasoning, but they also provide the mathematical proofs of such approach correctness.

In this paper the notion of centre-based bicluster is introduced. It is assumed, that bicluster values gather around the value called centre but the maximal difference between values and centre does not exceed an assumed level.

The paper is organized as follows: it starts from the short description of boolean reasoning for similarity biclustering of continuous data, afterwards the theory of centre-based biclustering is presented, next section presents the application of such approach for artificial data and the paper ends with some conclusions and perspectives for further works.

2   Boolean Reasoning in Biclustering

In the paper [8] the boolean reasoning approach for biclustering discrete and binary matrices was presented. For the given discrete value matrix the final boolean function was build upon all pairs of different cells. The pair of different cells was coded as the single disjunction clause, whose literals were corresponding to the cells row and column indices. The provided theorems claimed, that all prime implicants of such defined function corresponded to inclusion-maximal exact biclusters in the data. In terms of boolean context that approach was equivalent to transforming the Conjunction Normal Form (CNF) of the original boolean formula to the Disjunction Normal Form (DNF) containing only prime implicants.

For the binary matrices biclustering such approach was simplified: when biclusters of ones were searched, only the locations of zeros (meant as the logical sum of row and column indices variable) were used as elements of CNF for the further simplification.

Positive results of such approach were successfully extended for continuous matrices [9]. It occurred that it is possible to define the problem of similarity biclustering in terms of boolean reasoning. Again, the most significant element of matrix processing was to code all bicluster requirements violating cells as elementary clauses of CNF. That implied the new definition of bicluster of interests—centre-based—that are also defined and inducted in terms of boolean reasoning.

3   Centre-Based Biclustering

As it was presented above, it is possible to apply the boolean reasoning to find similarity biclusters of specified level of similarity. It is also possible to find all biclusters for all sensible levels of similarity analyzing only one boolean formula. But it may be interesting for data owner to find only such biclusters, whose values gather around a specified value with some margin of freedom. Let us call such bicluster a centre-based bicluster. Such bicluster is defined with two parameters: a centre c and the margin μ. This kind of bicluster will be now denoted as $b_{c,\mu}$.

Let us consider the continuous matrix M as presented in Table 1. It consists of five rows and four columns.

Table 1. Continuous matrix M_1.

	c_1	c_2	c_3	c_4
r_1	1	1	2	2
r_2	1	1	8	19
r_3	1	32	31	32
r_4	12	18	32	34
r_5	20	27	35	36

Let us assume that we are interested in finding biclusters whose elements gather around the value 30 with the margin equal 6. There are two of them that are inclusion-maximal and they are presented in Table 2.

Table 2. Continuous matrix M_1 with two $b_{30,6}$ biclusters.

	c_1	c_2	c_3	c_4
r_1	1	1	2	2
r_2	1	1	8	19
r_3	1	32	31	32
r_4	12	18	32	34
r_5	20	27	35	36

	c_1	c_2	c_3	c_4
r_1	1	1	2	2
r_2	1	1	8	19
r_3	1	32	31	32
r_4	12	18	32	34
r_5	20	27	35	36

It is obvious that the both biclusters fulfill assumed criteria for values (there is no cell in any bicluster whose value differs from 30 by more than 6) and for inclusion-maximality (there is no row or no column that could be added to any of biclusters without violating the criterion for values).

3.1   Boolean Reasoning Formalization

Let us consider the continuous matrix M and let us assume that we are looking for $b_{c,\mu}$ bicluster. Let us define the boolean formula of the form as follows:

$$f_{c,\mu} = \bigwedge (a \vee x)$$

where

$$a \in B, \quad x \in X$$

such that

$$|a(x) - c| > \mu$$

In other words we just code all cells whose values violate the similarity condition. The function $f_{30,6}$ for the matrix M_1 from the section above would take the following form:

$$f_{30,6} = (r_1 \vee u_1) \wedge (r_1 \vee u_2) \wedge (r_1 \vee u_3) \wedge (r_1 \vee u_4) \wedge (r_2 \vee u_1) \wedge (r_2 \vee u_2) \wedge (r_2 \vee u_3)$$
$$\wedge (r_2 \vee u_4) \wedge (r_3 \vee u_1) \wedge (r_4 \vee u_1) \wedge (r_4 \vee u_2) \wedge (r_5 \vee u_1)$$

To find the biclusters of interest the formula $f_{c,\mu}$ must be transformed from its CNF to DNF. Each prime implicant from DNF would then code one inclusion-maximal bicluster of specified condition. The CNF of $f_{30,6}$ is presented below:

$$f_{30,6} = c_1 \wedge r_4 \wedge r_1 \wedge r_2 \vee c_1 \wedge c_2 \wedge r_1 \wedge r_2 \vee c_1 \wedge c_2 \wedge c_3 \wedge c_4 \vee r_3 \wedge r_4 \wedge r_5 \wedge r_1 \wedge r_2$$

The formula below consists of four prime implicants. The implicant corresponds to the bicluster in such a way that bicluster consists of these rows (and columns) whose corresponding variables ARE NOT literals of the prime implicant. In Table 3 the detailed way of decoding all prime implicants is presented.

Table 3. Decoding the results of $f_{30,6}$ function DNF.

Prime	Literals		Bicluster			
Implicant	Rows	Columns	Rows	Columns	Bicluster	
$c_1 \wedge r_4 \wedge r_1 \wedge r_2$	r_1, r_2, r_4	c_1	r_3, r_5	c_2, c_3, c_4	$(\{3,5\}, \{2,3,4\})$	
$c_1 \wedge c_2 \wedge r_1 \wedge r_2$	r_1, r_2	c_1, c_2	r_3, r_4, r_5	c_3, c_4	$(\{3,4,5\}, \{3,4\})$	
$c_1 \wedge c_2 \wedge c_3 \wedge c_4$		$c_1, \ldots, c_4$	$r_1, \ldots, r_5$		$(\{1,2,3,4,5\}, \emptyset)$	
$r_3 \wedge r_4 \wedge r_5 \wedge r_1 \wedge r_2$	$r_1, \ldots, r_5$			$c_1, \ldots, c_4$	$(\emptyset, \{1,2,3,4\})$	

We may notice that two last biclusters are empty ones (Carthesian product of two sets, when one of them is empty, is also empty). But they also fulfill the mentioned requirements: there is no cell whose value differs from the 30 by more than 6 and such biclusters are inclusion-maximal. Empty bicluster of all rows and no column means that no column may become a proper bicluster (in terms of conditions). Accordingly, empty bicluster of all columns and no row means that no row may become a proper bicluster. We can confirm it that each row and each column of the matrix M_1 contains at least one cell whose value differs from 30 by more than 6. In the paper [8] the modification of boolean function definition for binary matrices was presented. It helped in some cases to avoid empty bicluster induction. It seems to be possible to apply simmilar approach in centre-based bicluster induction.

3.2   Proofs

Correctness (biclusters fulfill conditions) and maximality (biclusters are inclusion-maximal) of this approach is implied from two theorems, that are presented and proved below. Because there is the connection between logical variables and bicluster elements it is worth to explain the used notation. The row (and column) correspond boolean variable is denoted with the same notion. It simplifies the boolean formula, but it also requires to interpret r_1 as the first row or as a boolean variable, depending of the context where it is used. The set $\mathbb{X}$ is a set of row corresponding variables while $\mathbb{B}$ is the set of column corresponding variable. The set X is always a subset of $\mathbb{X}$ and set B is always a subset if $\mathbb{B}$. It is also necessary to define the $'$ operator. If X is the set of literals coding rows then X' is the subset of rows that are not coded by literals in X. Accordingly, if B is the set of literals coding colunms, then B' is the subset if columns that are not coded by literals in B. The implicant of boolean function is denoted as XB. The corresponding bicluster in the data is then denoted as $X'B'$.

Theorem 1 (Theorem of correctness). XB is an implicant of the $f_{c,\mu}$ function of the continuous matrix M iff $X'B'$ is a (c, μ) bicluster in M.

Proof. ($\Rightarrow$) Let XB be an implicant of the $f_{c,\mu}$ function and $X'B'$ be not the (c, μ) bicluster in M. That means that there exists at least one cell in the matrix M covered by the bicluster $X'B'$ whose value differs from c by more than μ. This leads to the statement that XB can not be the implicant of $f_{c,\mu}$ as it was assumed that $f_{c,\mu}$ covers all such cells in the matrix M.

($\Leftarrow$) Let $X'B'$ be a (c, μ) bicluster in the matrix M and XB be not an implicant of $f_{c,\mu}$. In such case it means that there exists at least one cell in the bicluster $X'B'$ coded by the implicant XB whose value differs from the c by more than μ, co the bicluster $X'B'$ can not be the (c, μ) bicluster.

Theorem 2 (Theorem of maximality). XB is a prime implicant of the $f_{c,\mu}$ function of the continuous matrix M iff $X'B'$ is an inclusion-maximal (c, μ) bicluster in M.

Proof. ($\Rightarrow$) Let XB be the prime implicant of $f_{c,\mu}$ and $X'B'$ be not the inclusion-maximal (c, μ) bicluster in M. We know from the previous theorem that $X'B'$ is (c, μ) bicluster of M. That means that there exists:

- $a \in \mathbb{B}$ such that $X'(B' \cup \{a\})'$ is (c, μ) bicluster or
- $x \in \mathbb{X}$ such that $(X' \cup \{x\})B'$ is (c, μ) bicluster.

From the same theorem one of the following:

- $X(B \setminus \{a\})'$ or
- $(X \setminus \{x\})B$

should be the implicant of $f_{c,\mu}$, which is in contradiction with the statement that XB is the prime implicant.

($\Leftarrow$) Let $X'B'$ be the inclusion-maximal (c, μ) bicluster in M and XB be not the prime implicant of $f_{c,\mu}$. As it was already proved XB is and implicant of $f_{c,\mu}$. Because XB is not a prime one that means that exists at least one of literals $x \in \mathbb{X}$ or $a \in \mathbb{B}$ that can be removed from implicant. The shortened implicant is equivalent (also from the theorem above) to the one of possible extended biclusters:

- $(X' \cup \{x\})B'$ or
- $X'(B' \cup \{a\})$

which is in contradiction to the statement that $X'B'$ is an inclusion-maximal bicluster.

4   Case Study

Let us consider a set similarity measure which is asymmetric. It can be defined as follows:

$$m(A, B) = \frac{|A \cap B|}{|B|}$$

It may be interpreted as the fraction of set B, explained by the items of set A. Now we consider the set of randomly generated subsets of the following set:

$$S = \{1, 2, 3, 4, 5, 6, 7, 8, 9, \ldots, 30\}$$

The set of random subsets is presented in Table 4.

In Fig. 3 the matrix of similarities between subsets is presented (M_S). The matrix is not symmetric because the introduced measure is asymmetric.

The level of similarity from the range $[0, 1]$ was mapped into the range in the gray scale $[0; 255]$, so the brighter pixel of the image in Fig. 3 the higher asymmetric similarity. In the presented case the goal was to find inclusion-maximal $(100, 50)$ biclusters. The total number of such biclusters was 915 (with two empty ones).

As it can be noticed in Fig. 4 most of the found biclusters have the mean value close to the assumed level 100 (the average mean value of bicluster is 91.58). The minimal value of minimal values in biclusters is 53.53 and the maximal value of maximal values is 100. That means that the condition of bicluster values range from the centre neighborhood is satisfied.

Moreover, found biclusters have rather wide area: mean value of the bicluster area is 64.8, top seven biclusters have the area equal 110 and 19 of all biclusters have the area greater or equal 100.

Table 4. Randomly generated subsets of S.

Number	Subset
1	{8,2,22,16,4,28,24,9}
2	{23,5,21,6,15,8,19,20}
3	{9,4,19,28,24,20,11,10,3,7 }
4	{27,13,16,10,12,6,3,28,9,4,1,21,5,18,26,30,24,22,7,29,19}
5	{16,15,24}
6	{30,8,12,3,13,14,7,1,27,11,22,19,21}
7	{24,7,28,17,10,26,19,4,8,14,3,1,30,2,23,29,13,15,22,9,27,20,16,18,11,21}
8	{8,24,18,1,17,20}
9	{24,26,28,10,2,23,12,14,25,17,7,18,30,27,5,20,4,15,29,11,1,13,22,16,8,3,21,19,6}
10	{7,24,13,19,18,1,9,11,28,29,14,30,16,15,2,4,5,12,10,6}
11	{15,9,16,1,25,10,3,11,2}
12	{30,16,12,8,4,10,15,28,5,11,20,14,24,17,3,21,1,25,27,13,29,6}
13	{30,18,27,1,24,2,29,6,20,17,9,28,13,23,15,7,10,3,14,11,25,8,21,19,12,16}
14	{24,29,18,11,19,20,4,28,21,30,25,8,16,3,12,7,27,1,23,26,2}
15	{10}
16	{9,13,29,12,20,17,10,1,21}
17	{15,24,28,20,14,18,13,17,21,11,19,27,7,22,29,23,2,1,5,4}
18	{12,9,14,26,2,25,20,8,21,6,18,17}
19	{27,19,3,24,12,13,18}
20	{11,9,12,29,17,14,10,8,21,19}
21	{18,29,9,22,27,10,8,6,15,13,4,16,30}
22	{26,8,20,1,19,9,18,16,30,5,12,4,13,11,14,24,22,25}
23	{11,30,18,10,13,8,29,20,7,27,5,1,12,17,14,19,16,2,23,9,25,21,6}
24	{27}
25	{21,8,3,15,20,6,19,24,7,18,5,2,30,28,14,23}
26	{23,20,22,3,4,5,25,28,9,2,18,19,16,15,24,29}
27	{10,8,20,15,12,26,29,18,3,1,21,14,2,7,23,22,17,27,6,19,5,13,24,11,4,25}
28	{7,19,30,10,1,5,17,11,20,16,12,4,6,27,29,2,23,24,9,3,25,8}
29	{14,8,15,27,29,22,25,20,3,9,11,10,16,28,4,5}
30	{21}

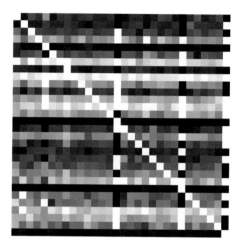

Fig. 3. Matrix M_S of similarities between subsets of S

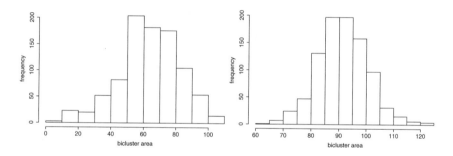

Fig. 4. Histograms of the bicluster area (left) and mean value (right).

5   Conclusions and Further Works

In the paper the new approach for finding biclusters in continuous data was presented. The novelty of the approach joins the boolean reasoning for biclustering and new definition of bicluster. The two aspects together brings the opportunity of performing a single step (building a proper formula for demanded biclusters) than inducting biclusters and postprocess them to find only those, that fulfill the value criterion.

The presented approach is strongly connected with the boolean function satisfiability problem. That causes, that the computational complexity of the algorithm is equal to the computational complexity of the 2-SAT problem (the CNF is the conjunction of two-literals disjunctions) which is polynomial.

For the purpose of calculation time decreasing, when only one (the top area) bicluster is needed, the modification of the Johnson's strategy [5], presented in [7], may be applied.

Presented experiments were performed in own developed environment, which will be published in the nearest future.

Acknowledgements. The work was carried out within the statutory research project of the Institute of Informatics, BK-204/RAU2/2019.

References

1. Brown, F.M.: Boolean Reasoning. Springer, New York (1990)
2. Busygin, S., Prokopyev, O., Pardalos, P.M.: Biclustering in data mining. Comput. Oper. Res. **35**(9), 2964–2987 (2008)
3. Chagoyen, M., Carmona-Saez, P., Shatkay, H., Carazo, J.M., Pascual-Montano, A.: Discovering semantic features in the literature: a foundation for building functional associations. BMC Bioinform. **7**(1), 41 (2006)
4. Hartigan, J.A.: Direct clustering of a data matrix. J. Am. Stat. Assoc. **67**(337), 123–129 (1972)
5. Johnson, D.: Approximation algorithms for combinational problems. J. Comput. Syst. Sci. **9**, 256–278 (1974)
6. Latkowski, R.: On decomposition for incomplete data. Fundam. Informaticae **54**, 1–16 (2003)
7. Michalak, M., Jaksik, R., Ślęzak, D.: Heuristic search of exact biclusters in binary data. Int. J. Appl. Math. Comput. Sci. (2019, to appear)
8. Michalak, M., Ślęzak, D.: Boolean representation for exact biclustering. Fundam. Informaticae **161**(3), 275–297 (2018)
9. Michalak, M., Ślęzak, D.: On Boolean representation of continuous data biclustering. Fundam. Informaticae **167**(3), 193–217 (2019)
10. Nguyen, H.S.: Approximate Boolean reasoning: foundations and applications in data mining. In: Transactions on Rough Sets V, pp. 334–506 (2006)
11. Orzechowski, P., Boryczko, K.: Text mining with hybrid biclustering algorithms. Lecture Notes in Computer Science, vol. 9693, pp. 102–113 (2016)
12. Tanay, A., Sharan, R., Shamir, R.: Biclustering algorithms: a survey. Handb. Comput. Mol. Biol. **9**(1–20), 122–124 (2005)

Issues on Performance of Reactive Programming in the Java Ecosystem with Persistent Data Sources

Lukasz Wycislik$^{(\boxtimes)}$ and Lukasz Ogorek

Institute of Informatics, Silesian University of Technology, Akademicka 16, 44-100 Gliwice, Poland
lwycislik@polsl.pl

Abstract. The paper presents research on the performance of reactive processing in the Java ecosystem where persistent data sources (both SQL and NoSQL) were used as one single dependency of the node. Several scenarios have been tested, including changing chunks of data stream, different types of database drivers, etc.

The results show that in the case of reactive processing of data streams being fetched from single noded persistent data sources it is hard to gain an advantage over nonreactive processing and some advantages began to appear only both with a large number of concurrent users and data streams of significant volume, while for low- and mid-loaded systems reactive processing gives usually unnecessary overhead resulting in degradation of the overall performance.

It should be also noticed that we are at the quite early stage of reactive programming development what results sometimes in lack of its support in various layers of technological stacks. This was the case with the PostgreSQL database, where there is no production release of the nonblocking driver, and no support for developing pipeline stored procedures yet.

Keywords: Efficiency · Java · MongoDB · PostgreSQL · Reactive programming · WebFlux

1 Introduction

The reactive processing in a field of computing is based on the assumption that data changes over time and flows in streams thus, may be processed asynchronously with operator-based logic. This allows processing data in a pipelined way what usually brings benefits in speed of computing. The concept of reactive processing together with modern trends of building software according to microservices architecture give the powerful tool for developing highly scalable, yet not wasteful, in terms of computing resources consumption, systems. Presently, the reactive processing approach is widely used for building many kinds of computing systems, e.g.:

- monitoring/control systems [3],

© Springer Nature Switzerland AG 2020
A. Gruca et al. (Eds.): ICMMI 2019, AISC 1061, pp. 249–258, 2020.
https://doi.org/10.1007/978-3-030-31964-9_24

- distributed computing systems [10],
- web applications [2],
- mobile apps [6].

Recognizing the advantages of reactive processing has resulted in the evolution of programming languages towards the so-called reactive programming that is based on introducing additional programming language abstraction to develop reactive applications, often but not necessarily, in a declarative way. This made the development of reactive systems even more productive and led to a situation when reactive programming is becoming a standard for building systems according to microservices architecture.

But is this approach a 'one-size-fits-all' solution bringing benefits in all cases of systems being built? Does the whole software development ecosystem is ready for reactive programming today? In this paper, the authors basing on conducted research, are trying to point out the scenarios where using the reactive programming approach may not necessarily bring benefits.

2   Background

An interesting review and attempt to categorize approaches to reactive programming shaped up to 2013 can be found in the article entitled "A survey on reactive programming" [1]. Already since then, apart from the benefits resulting from the use of reactive programming, many problems and dangers have been recognized.

One of the basic problems of reactive processing is making the system resistant to failures, and thus able to maintain internal consistency in the event of errors in various data processing points. This problem was noticed and discussed in more detail in [9], where authors argue that since none or very little effort has been undertaken to provide a single unified model able to tackle both event handling and shared state consistency, there is a need for developing such a framework, and they are proposing their own one. After that, there were also a few other proposals to mention among others ScalaLoci [13] or ReScala [8], which are being still actively developed.

In some cases, the possibility of mass scaling of data processing prevails over the need for a continuous consistency guarantee. Although dedicated application framwweworks are used to build such systems, their creators rarely show their practical advantage. The research containing such a verification were made in [7] on the example of the Lasp programming model.

Despite the fact that one of the main advantages of reactive programming consisting in the possibility of distributing of computations, and hence scaling to increase the overall system performance, constant work is underway to optimize the reactive processing platforms [4].

This constant development along with introducing reactive programming to a growing number of programming languages results in applying them more often as the default technology especially when microservices architecture is used.

3   Reactive Programming in Spring WebFlux

The concept of reactive programming has been popularized by ReactiveX.io – the reactive extension, which introduces reactive programming into languages, e.g. through RxJs in Javascript, RxJava in Java, Rx.NET in C# and many more.

In Java language there exist several reactive extensions where two of them presently are the most popular – RxJava2 and Reactor. A fair comparison of them was presented by Tomasz [12]. For the implementation of test cases used in this survey, the Reactor (Spring WebFlux) was used. It is included in the Spring Boot 2 that implements the new version of Spring Framework 5.

Reactor widely uses Java 8 API, e.g. CompletableFuture, Stream, or Duration. The reactor also brings two elements of reactive programming such as Flux (for handling n items) and Mono (for 0 or 1 item). The reactor obviously supports non-blocking communication between processes (IPC) and can use many application servers. It gives great flexibility and is very well suited for microservice environments. It also supports the backpressure mechanism, which gives the possibility of return communication between the producer and the consumer. If the consumer is not able to process the number of events produced by the producer, it will notify him that he can send the number of events that the consumer is able to process. The reactor uses the request mechanism here, informing the producer that at the moment it can process only n requests.

4   Survey

The survey was focused on a simple case where the node has one single dependency. This, of course, does not exhaust all possible scenarios for the implementation of micro-services, however, it is the most common [4] case and, according to the authors, the most interesting from the point of view of the expected results.

The benchmarks were carried out taking into account several dimensions. The first one was the persistent data sources where ones of the most popular open source databases were chosen – PostgreSQL 11, as an example of the relational database, and MongoDB 3, as an example of the NoSQL document database. The tests performed reading a data stream from databases, whereas in the case of PostgreSQL, the readout was implemented both with the SQL SELECT statement reading rows from the table and with the pipeline stored procedure using PL/pgSQL language. The last could be a common scenario in a case when one has to read data scattered on a complex relational data schema because when using declarative SELECT command there is limited influence on the execution plan developed by database management system engine, while when pipeline stored procedure is used the data can be accessed in an imperative way, what can favor implementing more efficient pipelined reading.

The second dimension was the volume of streamed data chunks. The tests performed for 1 KB, 10 KB, and 100 KB chunks. For PostgreSQL database, the chunk was a table row including one text column filled randomly with ASCII characters, while in the case of MongoDB the chunk was a JSON document including one text attribute filled in the same way.

The third dimesion was the number of data chunks in a stream to be processed as a single request. The tests performed for 10, 100, 1000 items. The fourth dimension was the type of data source drivers (synchronous vs asynchronous), while the last one was obviously the computing approach (blocking vs reactive).

4.1   Research Environment and Methods of Measurement

The research was conducted on the dedicated environment presented in Fig. 1 composed of t2.2xlarge EC2 instances (with "T2/T3 Unlimited" option enabled) set up in the AWS (Amazon Web Services). Each of instances consists of 8 vCPUs and 32 GiB RAM. The exact specification can be found in the AWS documentation pages [5].

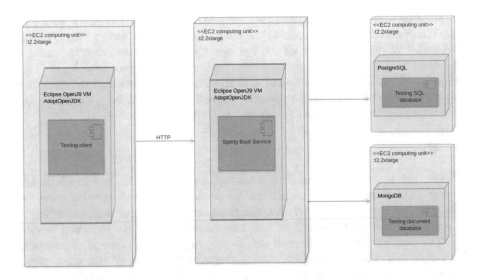

Fig. 1. The deployment diagram of the research environment.

As operating systems, Ubuntu 16.04 was used on all machines. Java applications were run using Eclipse OpenJ9 Virtual Machine and the network communication was configured using internal private interfaces.

For taking metrics, the Gatling, a powerful open-source load and performance testing tool for web applications, was used.

4.2   Measurement Methodology

All the measurements have been divided into two categories. The first one - several sequential reads being done by one virtual user. The results in this category give an answer regarding the maximum throughput of the system (understood as the mean shortest response times) for one user, therefore, they should be treated more as artificial indicators.

The second category includes tests of requests being done by many virtual users. Experiments cover scenarios for 100, 200 and 500 users, each repeating request in a time interval of a random value from 1 to 2 s. Before the maximum load is set, a certain time interval is provided for the gradual build-up of the full load (ascending ramp) and before the end, the same time interval was provided to suppress the load (descending ramp).

4.3 Results

The results are visualized as bar graphs where each bar shows the mean time of several system responses for a given scenario defined by dimension (a data source, a volume of the data chunk, a volume of the data stream, a data source driver, and a computing approach) and scenario category.

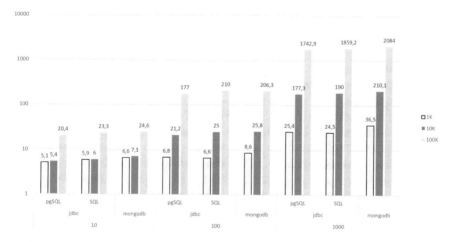

Fig. 2. Mean response times [ms] of MVC sequential requests for PostgreSQL and MongoDB.

First, the sequential approach for the MVC computing model was measured and results are presented in Fig. 2.

Comparing the results for the PostgreSQL database, it can be seen that fetching data using a stored procedure is almost always a bit faster than using standard SQL. This relationship is more evident for larger chunks of data, but rarely exceeds 10%.

MongoDB turned out to be almost always slower than the PostgreSQL database, what may be a bit surprising to the general opinion that NoSQL databases should generally perform faster. It is worth noting, however, that the test scenarios consisted only of reading data, and the configuration of each

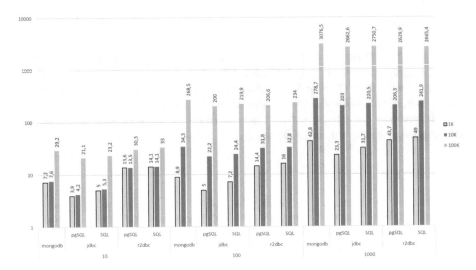

Fig. 3. Mean response times [ms] of WebFlux sequential requests for PostgreSQL and MongoDB.

database was composed of only one node. Thus, it did not reveal the advantage resulting from the easier scaling of the MongoDB database and simplified transaction management model.

The corresponding results for the WebFlux computing model are presented in Fig. 3. In the case of the PostgreSQL database, these tests were carried out for both the standard (jdbc) and the reactive (r2dbc) versions of the database driver. This was caused by the fact that the reactive version is for the time being an experimental version not recommended for use in production environments.

As in the previous case, it can be seen here that PostgreSQL is slightly ahead of MongoDB. It is more surprising, however, that the use of the reactive driver for handling WebFlux requests for the PostgresSQL database brought only performance degradation compared to the standard driver. This degradation is particularly evident in the case of small chunks of data.

In Fig. 4 the comparison of the MVC and WebFlux approach both to the PostgreSQL (standard SQL) and the MongoDB databases was presented. Only in one case the advantage of the WebFlux processing can be observed - this is the case of PostgreSQL for the lenght of the stream of data of 10 items with the use of jdbc driver. In most of the other cases the performance degradation can be seen, especially for smaller data chunks.

Next, the measurements for request scenarios of concurrent users are presented. Because this approach results in a high system load and significant memory consumption, tests were performed for the 1 KB of data chunks. In Fig. 5 the mean response times in miliseconds of concurrent requests implemented according to MVC model for PostgreSQL are presented. At first glance, it can be seen that even in the case of parallel processing, the use of the WebFlux model for the

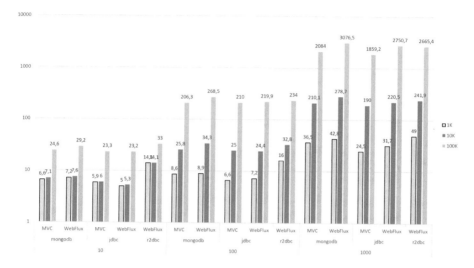

Fig. 4. The comparison of MVC and WebFlux for sequential requests to PostgreSQL and MongoDB.

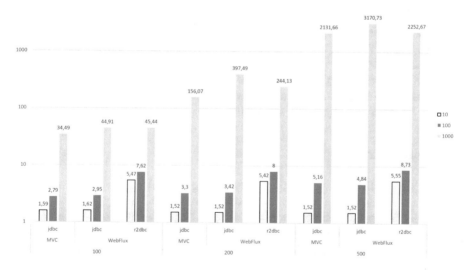

Fig. 5. Mean response times [ms] of MVC concurrent requests for PostgreSQL.

PostgreSQL database does not bring any benefits. It can be noted here, however, previously not observed phenomenon - the use of the reactive driver (r2dbc), for processing based on the WebFlux model, begins to bring benefits along with the increase of the data stream, in comparison to the use of the WebFlux model and the jdbc driver. The second observation is that in the case of increasing data stream, the reactive processing starts to outperform the MVC model (using the JDBC driver for shorter data streams and usign the r2dbc driver for longer ones).

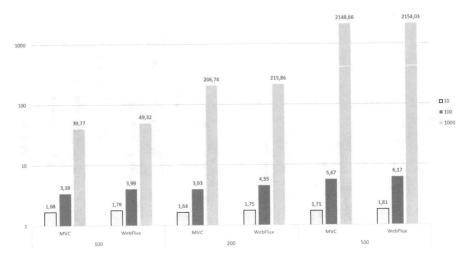

Fig. 6. Mean response times [ms] of MVC concurrent requests for MongoDB.

Analogous measurements for the MongoDB database are shown in Fig. 6, but here no improvement in performance can be seen in the case of using reactive processing. Figure 7 presents results both for MongoDB and PostgreSQL to direct comparison of these two database management systems. It can be observed that in the case of processing according to MVC model, PostgreSQL database always outperforms MongoDB. But in the case of processing according to WebFlux model, this is not always true, especially in the case of longer data streams.

5   Summary

In the age of distributing computation among several nodes according to the microservices paradigm, reactive processing seems to be a very suitable supplement that further enables horizontal scaling of computer systems. On numerous software development platforms, there can be found many artificial examples demonstrating the advantage of the reactive processing approach when the load is sufficiently heavy. However, real computer systems often implement single node dependency architecture and store data using database management systems. For such architectures, the performance study was carried out in this paper. It turns out that in such cases there must be special circumstances that the use of the reactive approach has a positive impact on systems performance.

On the other hand, however, the adaptation of reactive processing to the current technologies of microservices is still under active development. During the research described in this paper, the reactive database driver for PostgreSQL was still in the experimental stage and has not yet implemented the connection pooling what could explain some performance degradation over jdbc driver especially for short streams of data.

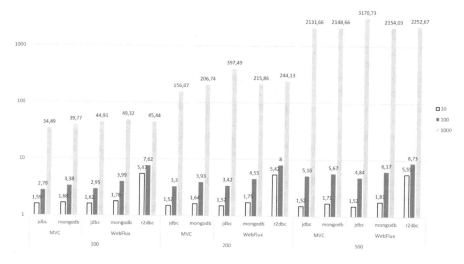

Fig. 7. The comparison of PostgreSQL and MongoDB for concurrent MVC and WebFlux requests.

Also, according to documentation [11], the PostgreSQL 11 database management system doesn't implement so far stored procedures in a way they could be pipelined: "The current implementation of RETURN NEXT and RETURN QUERY stores the entire result set before returning from the function, as discussed above. That means that if a PL/pgSQL function produces a very large result set, performance might be poor: data will be written to disk to avoid memory exhaustion, but the function itself will not return until the entire result set has been generated."

All of the above, and additionally the quite steep learning curve of programming languages constructions for reactive processing suggest that this architectural direction is not a 'one-size-fits-all' solution and should be chosen for system development when there do exist some rationales.

Acknowledgements. This work was supported by Statutory Research funds of Institute of Informatics, Silesian University of Technology, Gliwice, Poland (BK/204/RAU2/2019).

References

1. Bainomugisha, E., Carreton, A.L., Cutsem, T.V., Mostinckx, S., Meuter, W.D.: A survey on reactive programming. ACM Comput. Surv. **45**(4), 52:1–52:34 (2013). https://doi.org/10.1145/2501654.2501666. http://doi.acm.org/10.1145/2501654.2501666
2. Bernhardt, M.: Reactive Web Applications: Covers Play, Akka, and Reactive Streams, 1st edn. Manning Publications Co., Greenwich (2016)

3. Davari, M., Bertino, E.: Reactive access control systems. In: Proceedings of the 23nd ACM on Symposium on Access Control Models and Technologies, SACMAT 2018, pp. 205–207. ACM, New York (2018). https://doi.org/10.1145/3205977. 3208947. http://doi.acm.org/10.1145/3205977.3208947
4. Drechsler, J., Salvaneschi, G.: Optimizing distributed rescala (2014)
5. Jeff, B.: New t2.xlarge and t2.2xlarge instances. https://aws.amazon.com/blogs/ aws/new-t2-xlarge-and-t2-2xlarge-instances/. Accessed 19 Mar 2019
6. Jovanovic, Z., Bacevic, R., Markovic, R., Randjic, S.: Android application for observing data streams from built-in sensors using Rxjava. In: 2015 23rd Telecommunications Forum Telfor (TELFOR), pp. 918–921 (2015). https://doi.org/10. 1109/TELFOR.2015.7377615
7. Meiklejohn, C.S., Enes, V., Yoo, J., Baquero, C., Van Roy, P., Bieniusa, A.: Practical evaluation of the lasp programming model at large scale: an experience report. In: Proceedings of the 19th International Symposium on Principles and Practice of Declarative Programming, PPDP 2017, pp. 109–114. ACM, New York (2017). https://doi.org/10.1145/3131851.3131862. http://doi.acm.org/ 10.1145/3131851.3131862
8. Mogk, R., Salvaneschi, G., Mezini, M.: Reactive programming experience with rescala. In: Conference Companion of the 2nd International Conference on Art, Science, and Engineering of Programming, Programming 2018 Companion, pp. 105– 112. ACM, New York (2018). https://doi.org/10.1145/3191697.3214337. http:// doi.acm.org/10.1145/3191697.3214337
9. Myter, F., Coppieters, T., Scholliers, C., De Meuter, W.: I now pronounce you reactive and consistent: handling distributed and replicated state in reactive programming. In: Proceedings of the 3rd International Workshop on Reactive and Event-Based Languages and Systems, REBLS 2016, pp. 1–8. ACM, New York (2016). https://doi.org/10.1145/3001929.3001930. http://doi.acm.org/ 10.1145/3001929.3001930
10. Myter, F., Scholliers, C., Meuter, W.D.: Distributed reactive programming for reactive distributed systems. CoRR **abs/1902.00524** (2019). http://arxiv.org/ abs/1902.00524
11. PostgreSQL.   https://www.postgresql.org/docs/11/plpgsql-control-structures. html. Accessed 19 Mar 2019
12. Tomasz, N.: Rxjava vs reactor. https://www.nurkiewicz.com/2019/02/rxjava-vs-reactor.html. Accessed 19 Mar 2019
13. Weisenburger, P., Köhler, M., Salvaneschi, G.: Distributed system development with scalaloci. Proc. ACM Program. Lang. **2**(OOPSLA), 129:1–129:30 (2018). https://doi.org/10.1145/3276499. http://doi.acm.org/10.1145/3276499

Author Index

© Springer Nature Switzerland AG 2020
A. Gruca et al. (Eds.): ICMMI 2019, AISC 1061, pp. 259–260, 2020.
https://doi.org/10.1007/978-3-030-31964-9

Printed in the United States
By Bookmasters